COUNTRY LIFE BOOK OF

ANTIQUE
MAPS

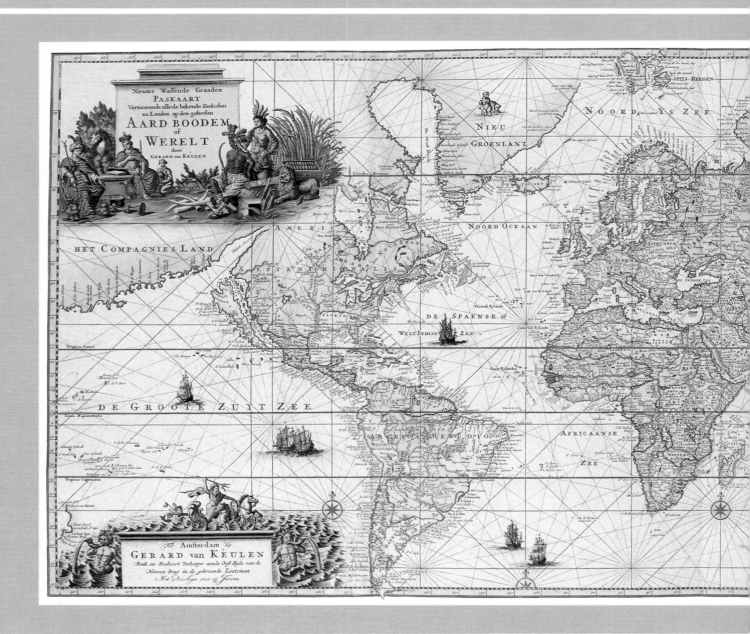

COUNTRY LIFE BOOK OF

ANTIQUE MAPS

An introduction to the history of maps
and how to appreciate them

JONATHAN POTTER

COUNTRY LIFE BOOKS

Preface

As first a collector of antique maps and now a professional map dealer of almost twenty years' experience, I have been aware in recent years of the particularly small number of books which introduce the subject of collecting and appreciating old maps.

Over the last thirty or forty years, numerous scholarly and carto-bibliographic works have been produced concentrating on specific aspects of the subject. However, these works are not appropriate reading matter for the person who has no previous knowledge of maps. Hopefully, this book will stimulate the reader's interest and lead on to more scholarly works in a chosen area of study.

Whether for an established map collector, a connoisseur of fine arts, or a buyer of decorative pieces, the choice of original maps available is surprisingly large in variety and diverse in content, appearance and price.

One of the first questions about old maps is 'Surely they are very expensive?' This can be the case, however given that many early maps can be purchased from twenty or thirty pounds upwards, there are a vast number of maps available for the purchaser in almost all fields of collecting and at all levels of the market.

Alternatively, fine World maps, for instance of the seventeenth century, may fetch a few thousand pounds but this represents remarkable value compared with many other fields of 'collectables', for few other areas can offer the same combination of scarcity, age, inherent interest, and decorative value.

I make no claims to this being a 'comprehensive' reference book; I hope, nonetheless, that the knowledgable collector will not only appreciate the selection of maps illustrated here but will also share the enthusiasm with which I have learned to look at maps. A book of this size cannot, of course, be a complete handbook, but I have tried to approach each aspect and area of map appreciation as if guiding an inexperienced collector around my gallery.

My purpose is to show something of the variety of original maps available on the market today and to provide an introduction to the wide world of antique maps – what to collect, how to collect, and, simply, how to enjoy them. When Edmund Clerihew Bentley wrote earlier this century in *Biography for Beginners*:

'The art of Biography is different from Geography. Geography is about maps, but Biography is about chaps.' ...

he could not have been thinking about maps such as those illustrated in this book. As I hope will be seen, 'Geography is about maps' but it also tells the stories of chaps who ventured where none had been before and lived to tell the tale.

First published in 1988 by
Country Life Books, an imprint of
The Hamlyn Publishing Group Limited,
a division of The Octopus Publishing Group,
Michelin House, 81 Fulham Road,
London SW3 6RB.

Copyright © Jonathan Potter 1988

ISBN 0 600 55626 3
Printed in Italy

ACKNOWLEDGEMENTS
My thanks are due to: Martayan Lan Inc. of New York for permission to reproduce the map illustrated on page 36, and to J. A. L. Franks Ltd, London for the illustrations on pages 51 and 179.
The great majority of photographs were taken by A. C. Cooper Ltd, London.

NOTE ON MAP MEASUREMENTS
Sizes of maps illustrated are approximate and are given width by height.

HALF-TITLE:
From Atlas (see page 41)

TITLE PAGE:
Nieuwe Wassende Graaden Paskaart ... Aard Boodem of Werelt by Gerard Van Keulen (see page 62)

Contents

INTRODUCTION

Over the last five hundred years the science of cartography has journeyed hand in hand with the development of the modern world. Maps, whether on flat pieces of paper or circular globes, have stimulated, assisted and recorded man's endeavours as profound as exploring the other half of the World or reaching the Moon, or as elementary and eminently practical as getting from London to Southampton or York.

Besides the prime geographical purpose for which the map is an essential document for identifying the known locations of places in relation to each other, maps have formed an integral part of every territorial debate, whether peaceful or otherwise, and have recorded the consequent national and social effects for all time. A map is a picture of the area shown. The sophistication of a modern Ordnance Survey map or hydrographical chart can provide the user with a mass of information: not only the safest way into harbour or the quickest route from A to B, but also the appearance of the country or coastline and its landmarks. The development of modern map making is a compendium of men's abilities and experiences, providing vivid records of that period of history when the Cape was rounded, America and Australia were discovered and the 'Unknown Parts' on the map of the World became known. Throughout this time map printing and production were practised as an art as much as a science and the consequent results are as enjoyable to study now as they were essential to man's knowledge in the years in which they were published.

Modern maps and atlases are essential to everyday life but often taken for granted – the maps we shall be examining in this book were produced to educate, to satisfy the requirements of travellers and merchants, to record new lands and the exploits involved in reaching them, to flatter the jingoistic ego or simply to amuse.

As Robert Louis Stevenson observed in *Treasure Island*, maps are 'an inexhaustible fund of interest for any man with eyes to see or two pence worth of imagination to understand with'.

Left

A New Map of England, Scotland and Ireland by Robert Morden (see page 77)

Above
'Geography Bewitched ... England and Wales'

Engraved by Dighton, published by Bowles and Carver.
London
c 1780.
16¹/₂ x 21cm 6¹/₂ x 8¹/₂in

*Designed to amuse – then as now – this curiosity
typifies a style that was practised occasionally throughout
cartographic history and, particularly, at this period. The shape
of England and Wales is transformed to a national caricature.
Contemporary colour.*

LOOKING
AT
MAPS

COLLECTING ANTIQUE MAPS

Antique maps hold a unique fascination compared with other decorative
or collectable antiques – their combination of art and science produces
a visual feast more informative than any picture can ever be.

Map collecting has been of interest to historians since the Middle Ages, although it is only more recently that early maps have assumed importance in understanding the geographical knowledge of the period. Those maps available on the market nowadays date from 1477 when map printing began, to the present day, although most map dealers' stock ends at around 1850 when map production, in the most broad terms, became purely functional to the exclusion of decorative features. Maps, after this time, are collected by those interested primarily in historical, sociological or geographical development and these will be examined in due course.

From the sixteenth century, maps and atlases were acquired as essential components of any good library. John Dee, Geographer to Queen Elizabeth I, noted 'some to beautify their halls, parlours and chambers … liketh, loveth, getteth and useth maps, charts and geographical globes' and, during the next century, the famous diarist John Aubrey remarks on the interest of collectors who 'loveth to visit booksellers shops there to lye gaping on maps'. Samuel Pepys was a noted collector and lamented the loss of his Speed atlas in the Great Fire of 1666. Map collecting is not, therefore, a new pastime. Abraham Ortelius, map maker and publisher, and regarded as one of the founding figures in the development of map publishing, was himself a collector of maps from an academic rather than a professional point of view.

The majority of published maps were first issued in books, either in atlases (as we now call them), travel or history books, guide books, or particularly, during the latter half of the eighteenth century, in periodical journals. Those maps not issued in such bound volumes may have been published in any of three other forms: as loose sheets, as wall maps (usually mounted on canvas with rollers top and bottom) or as folding maps (probably dissected, laid on linen and folded within binders or a slip-case).

In any of the instances above the vulnerability of the map on paper is obvious. As printed items one must assume 'production runs', even in the fifteenth and sixteenth centuries, of quite large numbers – hundreds or thousands at least – but the number of copies of one particular map which might have survived will be a tiny proportion of the original output. Consider the wear and tear of daily use, the ravages of bookworm or vermin, catastrophes such as the Fire of London or the Lisbon earthquake (1755), not to mention other civil and national disasters as well as the inevitable neglect with which 'out-of-date' objects are treated. Over two, three or four hundred years, though, any item which has escaped most of these dangers might, hopefully, still be found in good condition, the majority of surviving maps having originally been protected and bound into books.

Imperfect volumes of atlases and other books are therefore the source for most of the maps available to collectors today. Antique dealers, antiquarian book dealers, print dealers and map dealers have, over the years, taken maps and engravings from books defaced, perhaps, by fire or water, and from bindings or text damaged in other ways, maybe where a previous owner has removed sections of the book for his own purpose. We have only to think of the everyday treatment of books in the absence of the photocopier to imagine the dangers which might occur.

Which way to go?

For a new collector the first question is invariably 'What should I collect?'. The majority of collectors concentrate on one geographical area, for example, where they were born, where they live, where they have lived, where they would like to live, or because of some other specific interest or, of course, a combination of any of these factors. However, there are numerous other areas of collecting which I list below and which are examined later in the book.

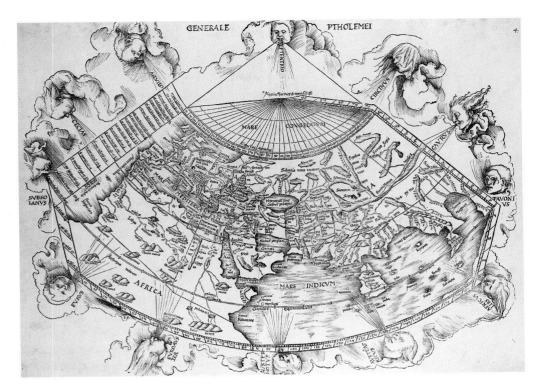

Left
Generale Ptholemei

Martin Waldseemüller's version of the Ptolemaic World Map.
Strassburg 1513
58 x 44.5cm 23 x 17.5in

The heavy lines of this wood-block engraving and its large size make this a particularly impressive version of the standard Ptolemaic depiction of the Ancient World.

Maps may be collected

1. By area.
2. By an individual map maker.
3. Of a particular period.
4. Showing particular geographical or decorative features.
5. By theme.
6. For their individual importance.
7. For their individual appearance.
8. For investment.
9. For any other reason the individual might find of interest.

1. *By area* This, the most obvious theme, allows the collector the greatest scope. He can set his own limits to the extent of his collection depending on his purse and time available. A comprehensive collection of major World maps prior to, say, 1800 could be a particularly expensive and long-term activity, whereas a collection of eighteenth-century World maps only would still include many decorative and interesting items at much less cost.

A collection of maps of a particular county, or region, whether England or elsewhere could be based on ten to a dozen maps relatively easily and inexpensively obtained. By concentrating on maps of one particular area, the geographical knowledge and social developments of that region can be clearly studied. A collector of English or European county or regional maps will record the development of villages into towns, name changes, growth of the road, canal and rail systems, boundary changes and so on; whereas, on a larger scale,

the collector of maps of Africa or Australia, for instance, can see firstly the continents' outlines coming into focus and gradual knowledge of the interior being plotted.

2. *By an individual map maker* Again, as above, there is enormous scope. The collector might choose any one of the great names in map making – Ptolemy, Munster, Ortelius, Mercator, Blaeu, Speed, Jansson, Sanson or de L'Isle, for instance. Each of these map makers has a distinctive style and the prices of their maps range, in most cases, from as little as fifty or sixty to thousands of pounds. In addition to this price range, there are also variations in content and format of an individual's atlases – hence maps by the Sansons can be found in three particularly different sizes; Ptolemy's *Geographia* was issued over at least 200 years in numerous editions differing in both size and style.

In the case of English map makers many, particularly during the eighteenth century, issued pocket-sized versions of larger atlases.

3. *Of a particular period* Styles and features of maps are frequently typical of a period of map making. Woodblock maps of the sixteenth century, for example, are distinctive. Well-engraved Dutch maps of the early part of the seventeenth century lacked the finesse of the later, more sophisticated productions; alternatively, the more precise, less flamboyant maps produced in England by Cary and Arrowsmith (c 1800) are characteristic of their period.

4. *Showing particular geographical or decorative features* A popular theme includes those maps showing bizarre or particularly curious cartography. California shown as an island is perhaps the most famous, and others

showing a series of seaways through North West America can be found. Decorative features on maps might include monsters or fabulous creatures, finely engraved ships or naval battle scenes, vignette views of native scenes, animals, birds and plants; panelled borders might include portraits of explorers or native characters in typical costume, town plans or panoramas and mythological scenes.

5. *By theme* Thematic collecting can incorporate any of a variety of particularly specialized interests – geological, railway and other 'social development' maps are products primarily of the nineteenth century, whereas road maps, historical maps, battle plans, cartographical curiosities and so on can be found from a much earlier period.

6. *For their individual importance* A collection could comprise maps which have various specific features: for example, the first map to show North and South America together by Munster, or the first map to show South America on its own by Gastaldi. Alternatively, maps showing Captain James Cook's voyages which mapped the Pacific – the last great undiscovered Ocean.

7. *For their individual appearance* Many maps merit admiration simply for their design and decoration. Although the map is the 'raison d'être' of the engraving, the artist could incorporate any amount of fanciful scrollwork, swags, rebuses, banners, ribbons, cornucopiae, cherubs, mythological figures, animals, birds, symbolic representations and scenes of the area shown within the design. The result in many instances is most decorative, particularly if the map is in good, ideally contemporary, hand colour.

8. *For investment* Although in the past – most notably in the late seventies – maps were pushed as investment commodities it must be stated that antique maps should not be viewed as ideal short-term investments. However, the long-term prospects for increasing map prices must be seen as very good. To quote figures would be pointless since no two copies of one map are ever identical – there will be differences in condition and colour – but there can be no doubt that with more and more people becoming seriously interested in the subject and with fewer maps coming onto the market each year, there will be price increases which will compare favourably with any other field of collectables.

The antique map market is very much in its infancy compared with stamps, silver, old master prints, etc, and as it develops, we must expect better quality maps of the more popular areas to appreciate at a greater rate than others.

If there are any 'rules of thumb' to follow in this area of map buying, one can only say to the investor 'buy what you like and at a sensible price'.

9. *For any other reason the individual might find of interest* The alternatives are infinite – maps on strange projections, maps in unusual languages (there are some in Armenian, Hebrew, Russian and Arabic besides most European languages), maps produced in particular places and so on.

How do I know what is genuine?

In the accepted terminology of map dealers and collectors, there are several words or expressions which can cause some confusion or question in the minds of the inexperienced. Amongst these, phrases such as 'original/genuine old/antique print/engraved map', 'a good copy', 'a fine impression', 'a late edition', etc spring to mind.

The honest dealer or map expert should be able to explain any of these terms – not least by being able to specify the date of publication of that particular map on that particular piece of paper. Many new collectors are surprised to find sixteenth- or seventeenth-century maps printed on strong paper and in good condition. 'Surely it is a later reprint?', 'How can it look so fresh?' they ask. The answers are that paper produced prior to around 1800 was generally of a much better quality than that produced more recently; that the life of a copper-engraved plate was relatively limited in terms of usage without being re-engraved; furthermore, the demand for an out-of-date map was negligible, and so the copper engraving would be scrapped, hammered out or melted down. Copper – the most commonly used medium for over 200 years – was an expensive material and, consequently, had to be reused for newly produced maps.

Many new collectors, wary in the knowledge that a number of decorative and sporting scenes are still being printed from nineteenth-century engravings, have initial fears that this must also apply to maps. Although a few of the original engraved plates survive, in my experience, none have been used to print maps commercially, in recent times. Simply from a commercial point of view, mass production of a rare map will depress the market, making the product unsaleable.

Unfortunately, there are no golden rules to guide the inexperienced map buyer. Experience and confidence will be gained only by seeing and handling the genuine item – this can be done more easily in the shops and galleries of map sellers than anywhere else. However, it is worth describing certain features here although they need to be seen on an old map to be fully appreciated.

1. *The paper* In the most general terms the majority of papers used for map printing were hand-made and of good quality. On close examination both visually and with the fingertips, the surface of the paper will probably appear very slightly rough or lightly textured, earlier paper tending to be more rough than that of a later period. When held up to the light a watermark

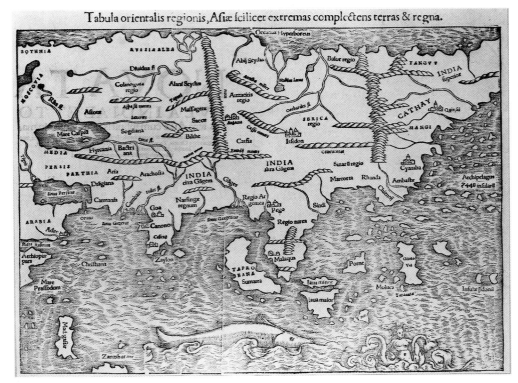

Right
Tabula Orientalis Regionis
...

Sebastian Munster's
woodblock of Asia – the
first of the continent in its
entirety.
Basle 1540
34 x 27cm 13.5 x 10.5in

Munster's issue of the 'Geographia' and, later, his compendium of geographical information, the 'Cosmographia' included the first maps of the individual continents.

(possibly) and chainlines (certainly) should be visible – these are the impressions left by the grid on which the layer of pulp was left to dry.

2. *The printed line itself* Depending on whether a woodblock or copperplate process of printing was used, there may be evidence of the 'impression' made by the printing block or its surfaces, which have actually come into contact with the paper. The 'blockmark' of the copperplate engraving will often be visible around the edge (the impression left by the edge of the copperplate after being pressed onto the paper).

An understanding of the two factors above should be able to confirm the information visually available on the map. For example, is the heavy, rather rough paper with strongly printed lines compatible with the geographical information, map maker's name and date engraved on the map? A little experience will soon build confidence.

The expression – 'a fine/poor copy' is often misunderstood – not surprisingly! For 'copy' use 'example'. The word 'copy' as used over the years is generally understood to mean reproduction; however, in this case, it applies to reproduction of the printing process in the course of a print run and not to reproduction of that engraving at a later date.

What is 'original'?

As with the word 'copy' confusion sometimes arises over the accepted definition of the term 'original', particularly when it is used to describe a subsequent reissue of a map.

With few exceptions engraved maps were published in more than one edition, or issue. In many instances

copperplates were reprinted over a period of time with alterations engraved on the map, for example, new geographical information was added or the publisher's imprint was altered; variations occurred regarding the book format (in size, or map location on a page of text); different language text was used. Sometimes at a later date the copperplate simply showed signs of age or wear.

In any of these instances the map is still 'original' although it may not be in its first edition. Thus, a 1676 edition of one of John Speed's maps is just as 'original' as the 1610–11 first edition of the same copperplate.

With regard to the age of the colour on a map, again experience is the surest guide. A map in 'early', 'original' or 'contemporary' colour (i.e., contemporary with publication of the map) may look surprisingly bright and fresh – early colour need not appear sombre or faded. As the work of various map makers is examined it will be apparent that many maps have a style of colouring which typifies the period. Ortelius, Hondius and Mercator, Blaeu and the English, Dutch, French and German schools of map making all used certain colours in clearly identifiable ways.

Maps, of course, that were issued in black and white – as virtually all until the nineteenth century were – may have been coloured at any time since publication. Clues as to the age of colour are few and the fact that there are now expert map colourists at work who can colour a map in the right style, with the correct colours, further complicates the issue. In most instances the acid base of the early verdigris green mixture used by colourists will show some 'see through' on the reverse of the map – in extreme instances the verdigris will bite through the paper and the whole area covered in that

green might drop out or disintegrate – rather unfortunate proof of 'original colour'. For a separate discussion
on the merits and disadvantages of 'modern' colour (see
page 19).

How many were printed?

This inevitable question is not one that is easily
answered. In only a few instances has research provided
estimates of original production. Unfortunately, publishing houses of the past either did not keep precise
records or, if they did, these have not survived for our
examination. However, it is acknowledged that a
copperplate would last, without the need for recutting
or re-engraving, for approximately 1000 'pulls' (i.e.,
impressions taken). Many generally more common
maps show evidence of this re-engraving process.

One might reasonably assume then that an average
production number for copperplate maps of the
seventeenth and eighteenth centuries would have been
in the region of some hundreds (in the case of known
rarities such as Gerard de Jode's maps) or as many as a
few thousand (estimated at between seven to eight
thousand in the case of some of Abraham Ortelius's
particularly popular maps, certain of which were
published between 1570 and 1612 in over forty editions
in seven different languages). Excluding separately
issued maps and considering only those bound into
atlases or books, it would be reasonable to assume that

even with the most popular atlases production would
not number more than a couple of thousand copies – of
which, of course, only a very small proportion might
have survived. Recent research suggests survival rates
of around eight to over twenty per cent for any of the
Ortelius maps, relatively few of which were published
in every edition of his atlas.

Given a survival rate as high as 20 per cent, the
actual number of any one particular map on the market,
or which might become available to collectors, is still
a tiny proportion of the original printing. Furthermore,
of these 'few hundred' that may still be extant, a great
number will be in libraries or public or private
collections, and will not appear on the open market.

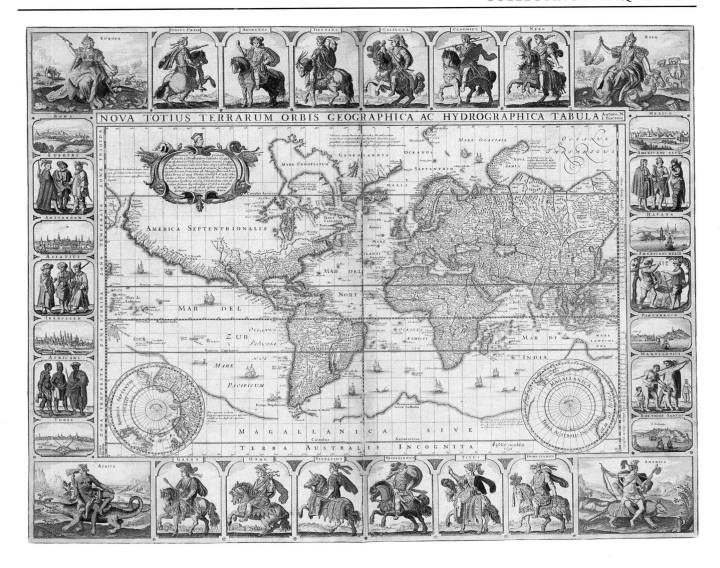

Above
Novia Totius Terrarum ...

Claes Janszoon Visscher's magnificent
World map on Mercator's projection.
Amsterdam 1639-52
56.5 x 45cm 22.25 x 17.75in

A particularly fine engraving. Twelve ceremonially attired Roman Emperors on horseback are flanked by corner depictions of each of the continents with native animals, while the side panels feature portraits of the inhabitants of Europe, Asia, Africa, North and South America and Magellanica and show views of Rome, Amsterdam, Jerusalem, Tunis, Mexico, Havana, Pernambuco and San Salvador. In original colour this map has some slight centrefold staining.

How much must I spend?

I don't believe there has ever been a map dealer or collector who has not, at some point, looked back in time at map prices and cursed himself for not spending or buying more in the past – this will always be the case. However, maps can still be found for a few pounds and good collectors' maps can be had from about twenty or thirty pounds upwards. Obviously for this price one should not expect great rarities of the most sought after areas; however, it should be possible to build up a large collection of rarities of a less popular area for not much more.

Unlike the relatively similar hobbies of stamp or coin collecting, there is no exhaustive catalogue summarizing the values of maps. This is not just because of the numbers of different maps that have been published – no more than the numbers of stamps – but because each map on the market has to be judged on unique factors, in particular, that map's condition and colour.

Condition factors to look for

Ideally, any map might be found in perfect condition. 'Perfect condition', 'mint', 'extra fine' and other superlatives would imply that the map is in exactly the same state as when published (i.e., with no damage and no repairs). However, unfortunately, this is not always the case.

15

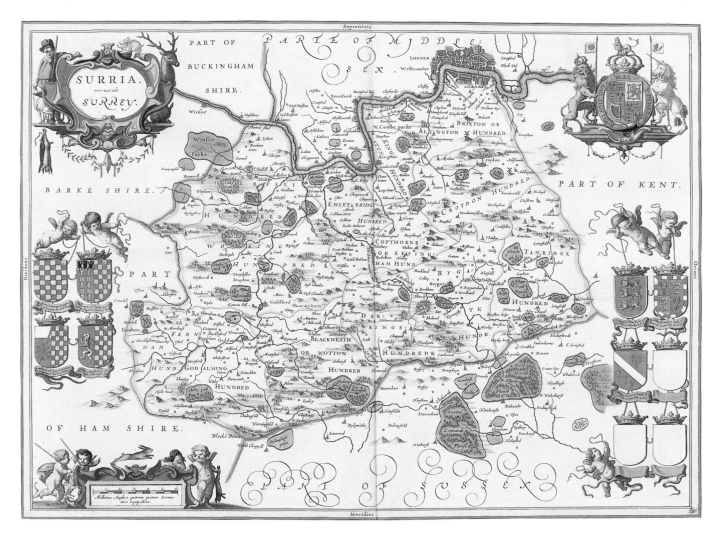

1. *Centrefold* One of the most common defects found in maps is a problem at the centrefold. The greater number of printed maps now on the market were issued in books and, invariably, spread across two pages rather than covering just one.

In the earliest publications 'stitchmarks' – where leaves have been sewn into the spine of the volume – may be encountered. This defect to maps pre-dating c 1520 is common and one must judge the degree of damage caused and the extent and quality of any subsequent repair. Although the woodblock maps of Sebastian Munster also often have this problem, the majority of maps issued during the remainder of the sixteenth century were bound into their volumes on 'guards' (for simplicity of printing and superior presentation the map, unlike the text, was printed on a double folio sheet; this was folded and then attached, by a tab, or 'guard' into the spine) and were better protected by this method. Nevertheless, the regular opening, flattening out and refolding of a volume of maps in day-to-day use frequently created wear at the centrefold, often resulting in a tear or split, normally at the bottom of the fold. This can be repaired and there should not be any loss of print. In extreme cases a split

Above
Surria ...Surrey

Jan Jansson's beautifully designed and engraved map of Surrey.
Amsterdam c1650
50 x 38cm 19.5 x 15in

Coats of arms of the nobility and of the Royal Family adorn this elegant engraving. The encircled green areas indicate parklands and the decoration around the title piece – a deer, rabbit, swan and gamekeeper give a clue to the genteel nature of this English county. This example is in good contemporary colour and on clean, untoned paper – not always the case with Jansson's maps.

might extend the length of the fold and careful examination is required to ensure that no actual print surface is missing.

2. *Tears* Tears, either as a consequence of a centrefold split, or on the edges of the map sheet may be encountered. Tears can normally be well repaired but, particularly if they extend onto the map's surface, they will devalue the map. Thinness, or weakness, within the actual fabric of the material may be found. Quite often an area of almost tissue-thin paper appears

within an otherwise perfect sheet of paper. Given the inconsistencies of the hand-made process, providing the face of the paper on which the map is printed is complete and intact, this is not a serious problem. My

---

Below
Nova Orbis Tabula ...

Frederic de Wit's fine double hemisphere
World map.
Amsterdam c1670
56 x 47.5cm 22 x 18.75in

A beautiful map which exemplifies the decorative qualities of Dutch maps of the period. Cartographically, the map's content is typical of the period – Abel Tasman's coastlines for New Zealand and Australia, one Great Lake and California as an island are shown. Decoration in each corner symbolizes the seasons and incorporates the signs of the Zodiac. An earlier state of this engraving lacks the outer border and cherubs within the cusps. This map is in original colour and shows early signs of verdigris deterioration.

---

suggestion would be to 'back' the thin area with archival material to guarantee the strength and stability of the map.

3. *Margins* Similar maps on different-sized sheets of paper may be found. This will normally be the result of the atlas or book, having been 'rebound' at some stage and 'trimmed' by the binder; alternatively, the map may have come from a 'de luxe' or 'special' edition of its atlas or book. The maps from Blaeu's *Atlas Major* or from early editions of John Speed's *Theatre ....* often have larger margins than those from other editions. Wider margins are preferable on maps although in many instances the margins are no larger than a half inch or so. The presence of the printer's 'blockmark' is important around the edge of the map, although occasionally a map may have been trimmed to the very edge of the printed area. Trimming close in this instance may have been done at the time of publication if the map was to be bound in a composite atlas. In such a case it is again necessary to make a judgement

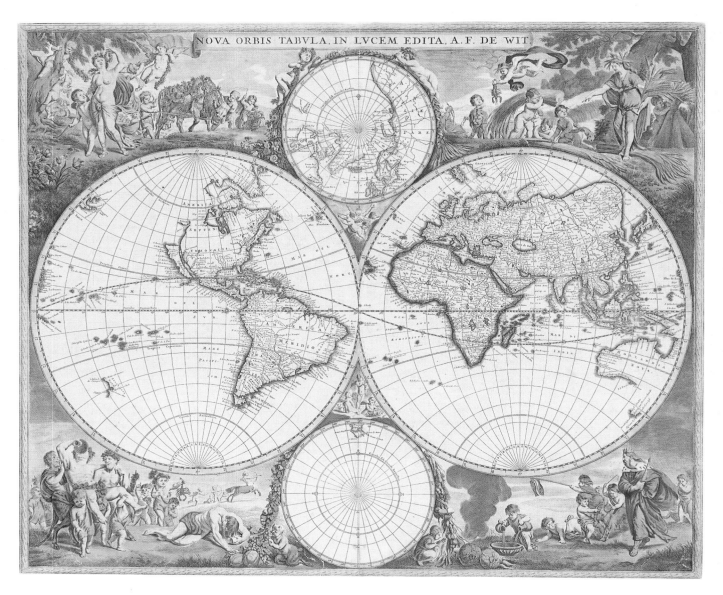

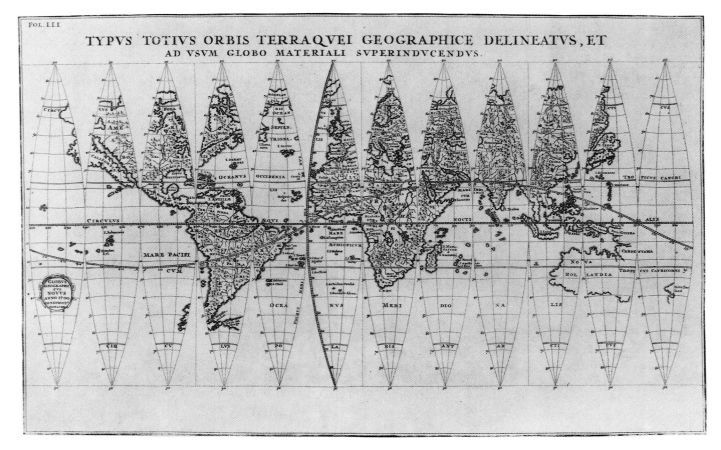

FOL. LII.

TYPVS TOTIVS ORBIS TERRAQVEI GEOGRAPHICE DELINEATVS, ET
AD VSVM GLOBO MATERIALI SVPERINDVCENDVS.

as to the effect of the defect on the map and the likelihood of finding the map in pristine condition, taking into account the consequent price differential.

4. *Wormholes* Maps of most periods may be found with small circular holes caused by bookworms. These small insects may have bitten away notable pieces of the page and although repairs can disguise the damage, any loss of printed detail may be seriously detrimental.

5. *Staining* Many atlases and maps have, at some time, suffered from staining of some nature. Usually water stains can be cleaned from the paper, but if there should be any oily content in the original staining liquid this may prove immovable. Tobacco stains and candle wax drops often leave permanent spotting. Yet again, the degree of disfiguration should be weighed against the price differential of a fine copy.

Problems related to colour

1. *Verdigris damage* The evidence of early colour on a map is often 'that the green shows through on the reverse of the map'. However, there are times when, either due to the original acid content of the verdigris mixture, or because the map or atlas has been kept in a damp atmosphere for some time, the areas coloured green have simply disintegrated. In the case of outline colour, the result may be that the area bounded by green simply 'falls out' of the map, or where a map area is covered in a green wash this will flake and disintegrate. If this problem is identified before any loss of surface,

Above
Typus Totius Orbis ...

Heinrich Scherer's set of globe gores.
Munich c1700
35 x 23cm 14 x 9in

Scherer's 'Atlas Novus' was an unusual publication – its emphasis being on maps of all areas of the known World, indicating their religious tendencies and, in particular, the extent of the Catholic faith and Jesuit Missions.

then backing the troubled area with archivist's tissue is the best remedy. Providing there is no loss of surface and any acidic activity has been halted the map should not be seriously devalued.

2. *Colour (and print) offset* This will either manifest itself as a faint mirror image transposed onto the opposite page or may be seen where the paint has attached itself to the opposite page and in some cases lifted the paper surface away with it. Occasionally, the colour and damaged attached surface can be lifted and replaced correctly. Rarely though can print offset be removed. This offset will probably be the result of binding the map and closing the book before the ink or colour on the sheet had properly dried, or because the book has been kept in a damp atmosphere and the moisture has reacted with the ink or colour, and made it spread to any facing fibres.

3. *Foxing* Smallish (penny-sized) brown spots, most commonly found on nineteenth-century paper, which are probably due to poor paper production and the subsequent effects of damp. Foxing can be removed by cleaning processes but these will also remove any colour on the map.

Modern v. old colour

One of the most frequently asked questions concerns the colouring of maps. 'Does new or modern colour on a map devalue it?' The answer is 'it depends'. The colouring of maps, as we have seen, may have been done soon after printing or publication. This is regarded as 'original' or 'contemporary' colouring. However, maps may have been coloured, or even recoloured, at any time since. There is nowadays a small 'cottage industry' actively involved in restoring and colouring old maps and prints.

In the broadest terms, recent colouring on a seventeenth- or eighteenth-century map does not reduce its value provided the colours used are in context with that particular map and have been properly applied. Recent colouring, which is not in the correct style for that map will devalue it. There is no doubt, strictly on commercial grounds, that a map, recently well coloured, will sell more readily as a decorative piece than the equivalent uncoloured item. However, a badly coloured example might prove unsaleable against the black and white equivalent.

Fine original colouring on a map in excellent condition is the ideal for any collector – maps satisfying these criteria certainly command a premium price. Unfortunately, relatively few maps are found of such a standard.

Personally, and without wishing to become involved in any ethical argument, my practice, as a dealer and collector, has been:

(a) to restore where necessary;
(b) to leave uncoloured those maps in good condition and in notably strong black and white impressions (invariably early 'pulls' from the copperplate); and
(c) when required, to have coloured correctly those maps which will visually benefit; where the quality of engraving will not be obscured; and, in general, to

Below
Les Isles Britanniques

Alexis Hubert Jaillot's boldly designed, decorative version of Nicolas Sanson's map of the British Isles.
Amsterdam c1710
89 x 57cm 35 x 22.5in

Generally regarded as one of the finest series of maps. Published first in Paris – then in Amsterdam, as in this case, by Pierre Mortier, in full original colour.

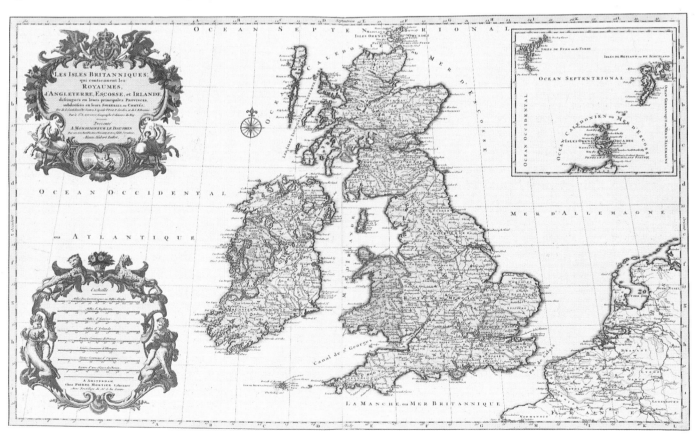

colour only those maps which one might reasonably expect to find 'originally' coloured (i.e., to colour an early Ptolemaic copperplate would not be appropriate whereas to colour a map by Blaeu or Ortelius, correctly, would be acceptable, given that these maps are often found in early colour).

Until the collector has sufficient experience to identify the age of colour himself he will be dependent on the dealer's expertise. Identifying the age of colour is not always easy but more often than not the experienced dealer can be confident in his opinion.

Below
Carte ... De La Caroline

Pierre Mortier's detailed chart.
Amsterdam c1700
59.5 x 48cm 22.25 x 19in

Based on an English map, this is one of the earliest detailed depictions of the country around Charleston. This example is in original wash colour.

As I have said there is undoubtedly a premium on fine original colour but, beyond that, colour applied, for instance, by a Victorian colourist to a seventeenth- or eighteenth-century map is certainly not modern. Equally, colour applied recently will soon no longer be modern. The final judge as to what is acceptable must be the collector and purchaser. Tradition has shown a place in the market for fine black and white maps though popular demand, through the centuries, has been for finely and correctly hand-coloured, visually decorative pieces.

The right price to pay – summary

To summarize this most important aspect of purchasing maps – until the beginner has sufficient experience to gauge the merits or defects of condition and colour, and the qualities of a certain edition of a particular map, he has to rely on the advice of the professionals. The dealer who knows his market should be able to provide comparable prices to support his asking price and thus reassure the uncertain purchaser.

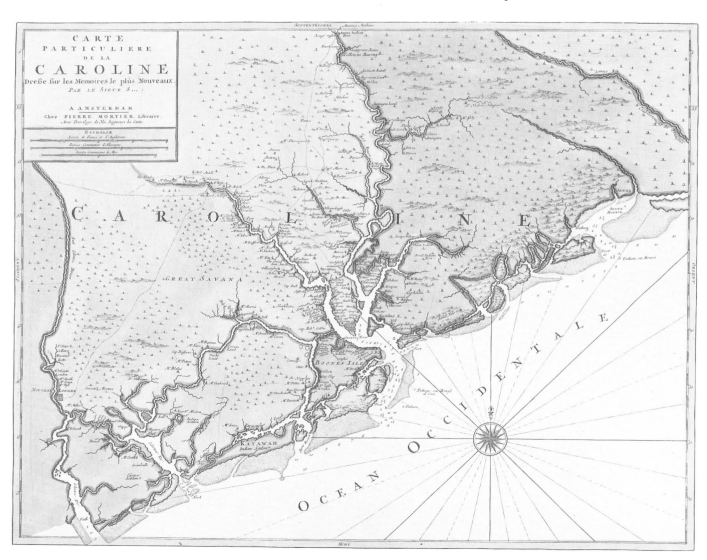

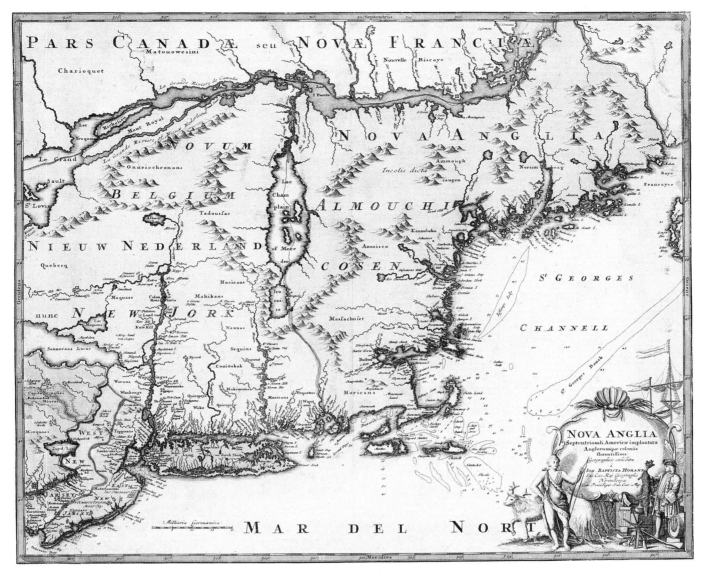

Above
Nova Anglia

Johann Baptist Homann's detailed map
of New England.
Nuremberg c1720
58 x 49cm 22.75 x 19.25in

A decorative map detailing European settlements along the coasts and rivers. Note the large, distorted outlines of Lake Champlain and three other lakes to the West of the map. The title piece shows a scene of a European trading with an Indian – furs, a barrel of liquor (or gunpowder) and weapons indicate something of the lifestyle of the period.

The pastel wash colour of this map is original and typical of German maps of this period, although it is often found in stronger tones.

The price of a map is the result of an equation which balances the following factors:

- By whom was it made? – an important or less important figure.
- What edition? – early or late, or important for any particular reason.
- What area is shown? – sought after or not.
- Is it in good condition or has it been repaired? What is the extent of the repair?
- Is the engraving a strong impression?
- Is the map well coloured, is it old colour?
- Is the map decorative?

Each of these factors is variable and also dependent on the prospective purchaser's likes and dislikes, but the map dealer will finalise a selling price on the above criteria linked to the cost of that map to him. Whether the purchaser should buy or not is his decision only, there is no specific price for one map – only a 'right' price.

LOOKING AT ANTIQUE MAPS

All who collect or just enjoy old maps do so for two fundamental reasons
– firstly, an immediate visual appreciation and secondly, an understanding of what the map shows and why it was produced.

Age and origin

'Where is the map of?' and 'How old is it?' are two
most common questions. Virtually all maps have a
formal titlepiece within their design, which will supply
the answer to this first question. Exceptions are the
Ptolemaic maps, where a title may lie outside the maps'
borders in some editions, and a small number of other,
generally less important, maps where there may be no
title at all. Newcomers to map appreciation are
sometimes confused by the fact that map titles are
frequently not at the top of a map. Such a convention,
commonly in use today, did not exist in time gone by,
and consequently the name at the top of a sheet will
invariably refer to the area above the subject area of
that map. Thus, for instance, the large lettered
'Hertfordshire' at the top of a Middlesex map refers to
the adjacent county and the title actually in question
may be found elsewhere on the map (see opposite).

The age of the map may not be self-evident. Dates
engraved on maps will generally refer to the date of
engraving or re-engraving of that plate, and not
necessarily to the date of publication of that map.
Frequently there was a delay during atlas preparation
between the commencement of the engraving of plates
and the completion of engraving and the actual
publication. Saxton's county atlas of 1579 has plates
with dates as early as 1574; some of Speed's maps with
the engraving date of 1610 were not altered for fifty or
sixty years. More often, however, maps did not have
dates or, after the first edition the date was erased from
the plate.

Why were maps undated? As will be apparent there
were few conventions in map production and dates
were not considered important. Furthermore, from a
commercial point of view the appearance of a date on
a map could limit the saleability of that item. The map
would appear outdated more quickly than its rivals. For
this reason, even today, dates are rarely found on
modern maps.

In the absence of an engraved date, how can a map
be dated? Few maps appeared without an author's,
engraver's or publisher's name on them. From any one,
or all three, of these details it should be possible to
identify the approximate or precise age of a map. The
engraver's name will usually be hardest to find and may
be 'lost' along the lower edge of the map or even hidden
within the decorative lines of the title cartouche. The
author and publisher were often one and the same,
although in the case of the maps from Bowen and
Kitchin's *Large English Atlas* (see page 87) the
combinations of the various publishers' names found
in the imprint below the map can serve to identify any
one of several editions exactly. A large number of
reference works are available to assist in identifying
specific editions.

How can the age be determined without a date and
a signature? Depending on the area shown, the
cartographic information itself may provide an answer.
For most areas outside Europe the information shown
may reflect the latest knowledge of an area; for example,
maps showing California as an Island (see page 160)
appear from about 1625 to around 1750; the name
'Australia' was not correctly used on maps until about
1815; and so on. In the case of British and European
county or provincial maps, more detailed knowledge is
required such as the dates of town, canal or railway
developments, or the changing of borders.

In the absence of clues to dating in the map's content
the style and presentation of the 'mystery' map should
be considered. In this case experience is the only true
guide to answering some of the following questions. Is
the paper thick, heavy or fibrous, or thin and smooth?
What is the method and style of printing? Is the map
printed from a woodblock, a copperplate, or lithographed? Is the quality of engraving heavy or delicate,
flamboyant or unfussy? Does the map bear notable
characteristics of any particular period or map maker?
Does the map look and 'feel' right for a suggested date?

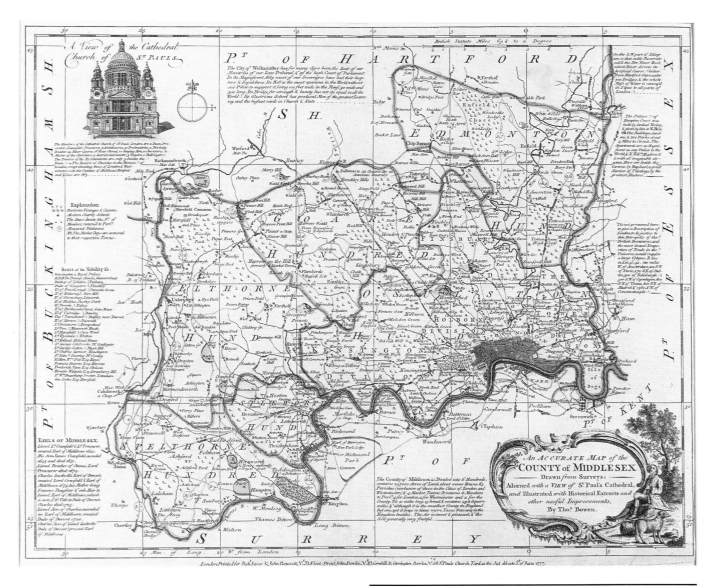

Author and purpose

Having established the area shown and the age of the map, one might question by whom and why the map was made.

The answer to the first of these questions will probably be self-evident on examination of the map, whereas, the answer to the second question may not be so apparent. All maps provide information for reference or practical use. Maps bound into atlases were most often lodged in the libraries of universities and colleges, monasteries and the private collections of royalty and landed gentry. Loose sheet maps and charts were more frequently for practical use, by travellers or maybe for military purposes.

The decoration and style of presentation of the map are indicative of the map's purpose or origin. Early engraving tends towards a heaviness or crudeness in the engraved line, which becomes more refined as the technique develops. Thus early woodblock engravings of Ptolemaic maps appear much 'heavier' than the later Munster woodcuts. Equally the copperplate engravings

Above
County of Middlesex

Thomas Bowen's map from
The Royal English Atlas.
London 1763 – this edition 1777
49 x 41cm 19.5 x 16in

An uncommon map which contains a wealth of cartographic detail with paragraphs and lists of information all around the map. On the map, which covers the greater part of today's London area, roads, rivers, villages, hilly areas, churches and parklands are shown; whilst the notes around describe the histories of London, Westminster and Hampton Court, and list the Noble Seats, the Earls and statistics of relevance to the county.

N.B. The words 'Pt. (i.e. Part) of Hartfordsh.' at the top are not the title of the map. The engraver's and publishers' names and date are all apparent.

The outline colour on this map is typical of the period.

of the so-called 'golden age of cartography' of the seventeenth century show much finer engraved lines than are apparent on the maps of Ortelius of the late

23

1500s. Conversely, however, eighteenth-century copperplate work reverts to a heavier, bolder line to satisfy a larger, but less discerning public.

Maps dedicated to royalty in the hope of patronage or advancement, or to sponsors or patrons of the arts and sciences will usually be more artistic productions than those merely illustrating a history or geography textbook.

The condition of the map might indicate if it was much used, and manuscript annotations might give some insight as to the original owner or user. A map in magnificent contemporary colour, for instance, might have been for presentation or for a particularly wealthy customer.

Map design and format

In broad terms little conformity existed in map and atlas production until the seventeenth century. Ortelius's atlas (first published in 1570 and generally regarded as the first atlas in present-day terms) was ultimately composed of maps of different scales by numerous different surveyors and engravers.

The map's orientation would be identified by means of a compass rose, varying in its degree of elaboration, or simply the words or forms thereof, for Septentrionalis (North), Meridionalis (South), Orientalis (East) and Occidentalis (West), placed adjacent to the map edges. Confusingly, the concept that North should appear at the top of the page was not always observed.

Map borders, initially a single straight line or parallel lines in which a scale of gradation would appear, evolved into elaborate strapwork, latticework, rope or floral designs.

Right
Novi Belgii

Nicolas Visscher's beautiful map and
picture of the United States' north-east
coast and New York.
Amsterdam c1660
55.5 x 46.5cm 22 x 18.5in

This is one of the best-known and most sought-after maps of the area. Based on Jan Jansson's map of 1651 and on a view of New York about the same time, the map details European settlement from the Chesapeake to Maine and contains the most up-to-date information with vignette illustrations of, amongst other creatures, beavers, turkeys, bears and foxes and scenes of native Indian villages. New York retains the original Dutch title of 'Nieuw Amsterdam' and is flanked by native Indians.

The map, and derivatives of it, appear in numerous different editions and forms – the series being known as 'Jansson-Visscher' maps – for about one hundred years. Identification of specific publication dates are possible by reference to specific authorities.

This example is in bright original colour.

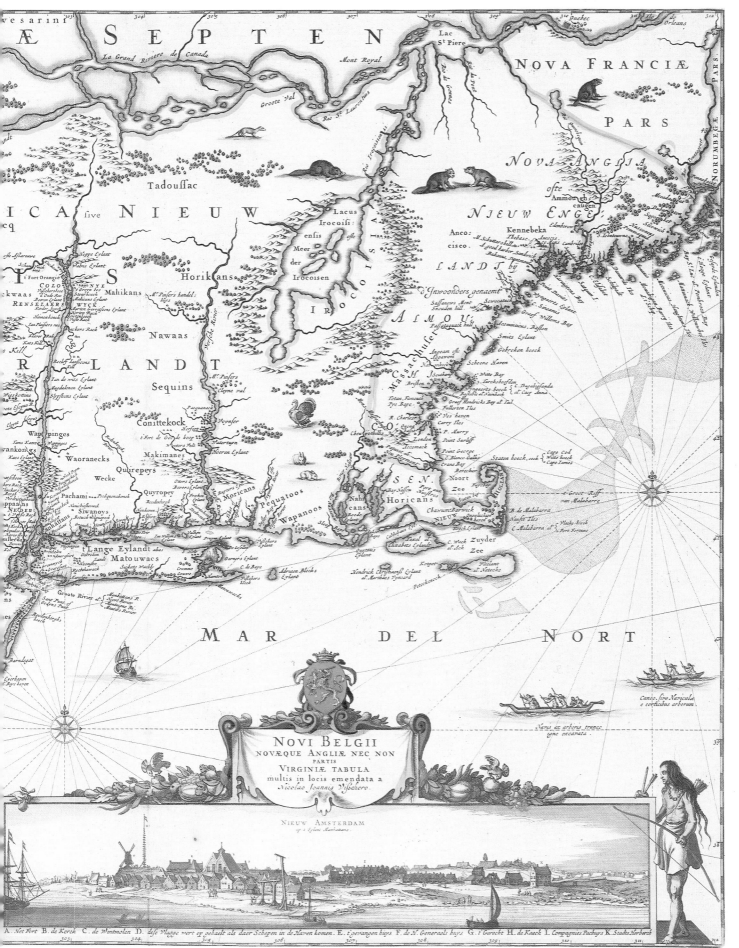

NOVI BELGII
NOVÆQUE ANGLIÆ NEC NON
PARTIS
VIRGINIÆ TABULA
multis in locis emendata a
Nicolao Joannis Vischero

The earliest maps – Ptolemaic and otherwise – were often devoid of any decoration or ornamentation either in cartographic presentation or additional features. As the interest in, and importance of, maps as printed documents developed, so the titles of maps were set in more ornate surrounds – cartouches. Hence, by the end of the sixteenth century we see intricate strapwork designs, the incorporation of gargoyles and fantastic creatures and other elements taken from the architecture and sculpture of the day.

Compass roses, mileage scales, keys to symbols, and panels of text enclosing author and publishing details, all provide an opportunity for the artist/engraver to demonstrate his talent.

Map decoration

During the seventeenth century the lavish embellishment and decoration of maps became the standard practice. Thus, we see the title cartouches being used to support illustrations of characters, and flora and fauna native to the area shown. Vignette views in blank areas of the map further illustrate scenes of local life or native economy. Through these illustrations it is possible for the map to convey a great deal of information regarding the inhabitants and nature of an area.

Very often these illustrations would be based on the reports of the earliest European travellers and would, as such, be the first pictures many Europeans would see of the Americas, Africa and Asia.

As will be apparent, an atlas was produced with a view to its saleability. Maps would be designed in more decorative and appealing ways but, above all, the atlas frontispiece and the World map would invariably be spectacular productions intended to stimulate the prospective atlas buyer and persuade him that within this volume would be found finer maps than in any other.

Map makers were not averse to stimulating the reader's curiosity by theorizing on geographical features in unknown parts or illustrating fantastic beasts in blank areas on the map. As Jonathan Swift remarked:

> 'So geographers, in Afric-maps,
> With savage-pictures fill their gaps;
> And o'er unhabitable downs,
> Place elephants for want of towns.'

Each period of cartography from the medieval maps after Ptolemy to the precision of the Victorian era has produced some wonderful and exciting maps in a multitude of visual styles, combinations of map maker's science, draughtsman's art and engraver's skill.

Cartes à Figures

By the start of the seventeenth century, the shift of prominence in map producing from Italy to the Low Countries was complete and the flamboyant, large-scale wall maps being produced of the world, and later, its continents by Plancius, Blaeu and others were crammed with detail in addition to their cartographic content. Outside of the various formats of decoration for World maps, the use of a regular, formal, panelled border on three or four sides of the map became most popular for folio size maps which were either issued separately or bound into atlas volumes.

Maps of this type, which became known as 'cartes à figures' are very rare – either as a result of being issued separately and simply not surviving until now, or since the engraving was often larger than average, they were trimmed when bound, or folded several times and are thus, if found on the market, in relatively poor condition.

Maps by Pieter van der Keere, Claes Janszoon Visscher, Guillielmus Blaeu, Jodocus Hondius and others may be found in the first half of the century and later by Frederic de Wit; whilst in England, John Speed, John Overton and Robert Walton produced similar versions of the work of their Dutch counterparts.

Curiously, the style is peculiar to the seventeenth century – presumably, since later atlases had detailed maps on each part of those continents previously shown in only broad detail on more general maps.

Invariably, panels at each side would show full-length portraits of the inhabitants of the area shown, and the top and, if present, lower borders would show panoramas and plans of that area's cities, ports and features of interest.

Further examples of this type of map are shown on pages 15 and 136.

Right (above)
Nova Hispaniae

A spectacular figured border map by Jan Jansson.
Amsterdam 1632
56 x 46cm 22 x 18in

In bright original colour this fine map was, in the first instance, separately issued and has survived, in this particular example, since it was folded and bound into a composite atlas of the period. Besides the full-length portraits and panoramas, the Spanish Coats of Arms and a portrait of King Philip III can be seen and, beneath the orange branch bedecked titlepiece, a view of Madrid.

Right (below)
Nova Europae Descriptio

An elegant, later bordered map by Frederic de Wit.
Amsterdam c1660
56 x 43cm 22 x 17in

De Wit's uncommon series of continent maps was replaced in his atlases (c 1670) by a less decorative, more detailed map almost lacking embellishment. Six of Europe's major cities are shown at the top – Rome, Amsterdam, Paris, London, Seville and Prague and the eight bordering panels show full-length portraits of the rulers of European countries. Like the majority of Dutch-produced atlases of the time, this example of de Wit's map is in contemporary colour.

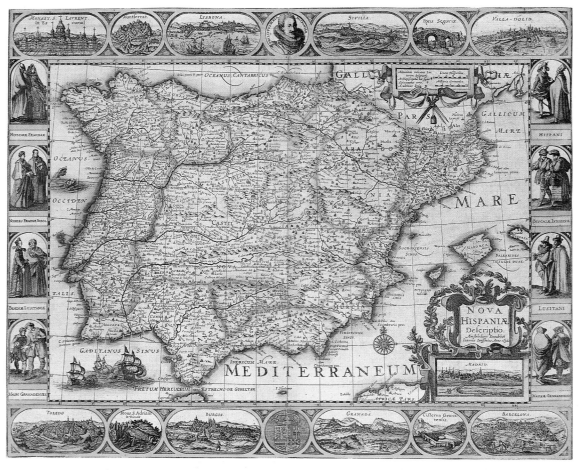

MAPS AND MAP MAKERS

In recent times, the collecting of atlas title pages and cartographers' portraits has gained popularity. Particularly decorative and well designed title pages have appeared in atlases throughout the centuries, whereas map makers' portraits are far less often found.

Even before a potential atlas buyer of the seventeenth century had seen the first map in a book he would have encountered the title. Often this was printed on two pages, firstly a decorative engraved title page and then a detailed letterpress description stating title, author, printer, publisher, place of publication, date and any other information which might whet the customer's appetite to look further.

Frequently, the engraved title would be the most ornate and finely worked item in the book. These titles were often designed and engraved by the foremost artists of the period whereas the maps, by their very nature, were relatively limited in imaginative scope. The actual design of the title page decoration varied and developed over the years.

The earliest forms of embellishment were simply ornate frames for the book title, very occasionally incorporating classical figures. This simple frame evolved in two distinct directions, each serving the same function. The simplest, in concept, is the strapwork design, seen in Mercator's three-part atlas titles and various others around that time. This strapwork style design was often used as a frame for map title panels.

The other extension of this 'framed' title theme puts the title lettering within an architectural setting, frequently with alcoves or pediments on which statuesque figures, symbolizing, perhaps, inhabitants of each of the continents or mythical characters, are positioned.

One of the best known designs of this style is that of Ortelius's *Theatrum Orbis Terrarum* title page. Either side of a colonnaded arch are female figures representing Africa and Asia, above is the gracious figure of Europe and seated in the foreground an Amazonian figure,

Spectandum dedit Ortelius mortalib. orbem,
Orbi spectandum Galleus Ortelium.

Left
Abraham Ortelius

The most influential character in atlas making.
Antwerp 1579 onwards
32 x 21.5cm 13 x 8.5in

Born in 1527, died 1598. This portrait is surrounded by an ornate strapwork and floral frame. Ortelius' interests in history, topography and cartography led him to the production in 1570 of the first comprehensive collection of maps of all parts of the world – the first atlas – as we now know such a volume.

holding a decapitated head representing the New World. Exceptionally, an explanation of the design is provided by a poem within the atlas preliminaries and the most curious part of the design, a female bust on a pedestal above a flame is described as representing the southern tip of South America, Magellanica, whose full extent was unknown and where fires (hence Tierra del Fuego) had been sighted by many sailors.

The *Theatrum* title is simple but striking. Numerous later title pages, however, are striking in their complexity of design and the wealth of their content. Some of these are magnificent examples of design and engraving and, if found in good original colour, are particularly decorative.

A variation on the theme of symbolism occurs
where the title is entirely secondary to design, the
magnificence of the engraving overwhelming the title.
Invariably a full page of letterpress accompanies the title
in such instances. Within this category are many
designs which incorporate surveyors' and navigators'
instruments, scenes of surveying and map making, or
of shipping and commerce.

Title pages with maps themselves in their design are
sometimes found – one of the most curious of these is
Theodore de Bry's issue of Herrera's voyages published
in 1624, which features a map of North and South
America with the earliest representation of California
as an island (see opposite).

In addition to the decorative title page, some atlases
included a portrait of their author. Occasionally,
portraits of the more celebrated map makers might
be engraved as separate publications or for inclusion
in album collections of the more important figures of
the day.

Consequently, it may be possible to find portraits of
the following European map makers – d'Anville, W J
Blaeu, de Bry, Coronelli, de L'Isle, du Val, Homann,
H Hondius, Linschoten, G Mercator, Munster, Ortelius, Ptolemy, or Sanson. British map makers who
might be found include Arrowsmith, Camden, Drayton, Hollar, Ogilby and Speed. Neither list is in any
way exhaustive, but they do merely illustrate the wide

Published by Fielding & Walker 1.st Feb 1780.

Above
Capt. James Cook

A delicate copperplate engraving.
London 1780
9.5 x 15cm 3.75 x 6in

Published within twelve months of Cook's murder, in February 1779, on Hawaii, this dignified portrait indicates Cook's fame as a geographer and navigator by way of the globe and telescope, and, in a scene below, records his murder and shows fleeing British mariners being chased by club-wielding natives.

Above
Descriptio Indiae Occidentalis

Theodore de Bry's title page for his edition of Herrera's accounts of the New World.
Frankfurt 1624
19 x 29cm 7.5 x 11.5in

This curious title page has circular portraits of Columbus and Vespucci and full-length portraits of Magellan and Pizarro – the latter pair with compasses and maps. These figures border, on a scroll, a map of North and South America, which shows, apparently, the first printed depiction of California as an island.

scope for the collector in this particular area.

In addition to the map makers, there are portraits of astronomers and scientists, for example, Brahe or Halley and explorers and navigators such as Columbus, Cook or Drake whose impact on the development of World cartography was, in many cases, immense.

LANDMARKS IN MAP MAKING

Throughout the development of commercial map publication it is possible to identify certain outstanding 'landmarks' or highlights which, in themselves, summarize cartographic progress through the centuries. Each of the maps specified have had a significant effect on subsequent map publication and in the following chapter it can be seen how European geographical knowledge of the time was reflected on paper.

In the beginning

The earliest printed map was a very simplistic diagram of the world now known as the 'T-O' form. Derived from medieval manuscripts the circular form consisted of an ocean round its circumference within which three landmasses, separated by a T-shaped waterway, were identified – Asia, Europe and Africa. The map was first published in 1472 in an encyclopedic style work originally compiled by a seventh-century bishop, Isidore of Seville. This work was printed in Augsburg within some twenty years of Johann Gutenberg's printing innovations which had culminated with the production of the first printed Bible.

Although the earliest editions of Claudius Ptolemy's *Geographia* were printed in Italy, (at Bologna in 1477, Rome in 1478 and Florence in 1482), the publishing houses of Central Europe were also producing some interesting maps in these very early years.

In 1482 the first edition of Ptolemy's *Geographia* using woodblock maps and produced north of the Alps was printed by Leinhart Holle. The World map in this atlas, although based on the traditional Ptolemaic style, showed progress in that it added a representation of Scandinavia outside the regular trapezoidal frame, and the first map devoted to this area was amongst the new maps added to this edition (see page 34). The atlas was

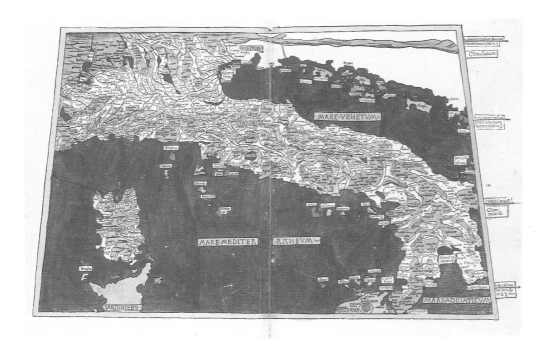

(see page 34)

Left
Untitled

An untitled map of Italy, the first atlas-producing country, from the first edition of Ptolemy's maps to be published outside Italy.
Ulm 1482
56 x 35 cm 22 x 14in

This strong original colour typifies the series. The second issue of this map, 1486, is also often in original colour which frequently differs in appearance, in that the sea will be of an 'olivey'-brown hue.

This particular map is one of the five 'modern maps' added to the standard Ptolemaic format by Germanus and published by Holle.

reissued four years later and it is the earliest series of maps which might be found in original colour. (See page opposite and pages 70 and 71 for the richness and depth of colour used).

The renowned *Nuremberg Chronicle*, published by Anton Koberger and edited by Hartmann Schedel, is one of the great books of the Renaissance and includes maps of the Ptolemaic world and Europe. The World map is undoubtably one of the most curious of any and is really the last pre-Columbian view of the World (see below). The book, published in 1493, is a compendium of historical and scientific fact, myths and legends and is illustrated with innumerable woodblock scenes and city panoramas (see page 108).

Not only did the last half of the fifteenth century see the science of printing develop dramatically, but also European knowledge of the world and its terrestrial form expanded in a cataclysmic way. During the century the Portuguese, under the influence of Prince Henry the Navigator, who died in 1460, had been pushing the bounds of knowledge of the African western coastline further and further southwards. Progress had been slow, despite Henry's efforts, due in part to the calls of military and naval campaigns, the distractions of commercial enterprise and, not least, the traditional

fear of Europeans for the Atlantic Ocean in southern latitudes. Despite the improvements, over the period, in ships and navigating techniques, (developed through Henry's school at Sagres, the most south-westerly point of the European mainland), the fear persisted that the Sun could turn men black, that the sea boiled, and that the ocean current accelerated to sweep ships over the edge of the world. Within twenty years, either side of the turn of the century, the map of the World was redrawn. This period of enlightenment effectively opened up the World to Europe and her influence, for better or worse.

Below
The Last of the Ancient Worlds

The curious map from Hartmann Schedel's
Nuremberg Chronicle.
Nuremberg 1493
52 x 38.5cm 20.5 x 15in whole sheet size

A famous wood-cut map featuring an outdated Ptolemaic world with a land-locked Indian Ocean with flanking portraits of Noah's sons – Japhet, Shem and Ham. The bizarre figures outside the map but on the same printed page add to the curious nature of this sought-after piece.

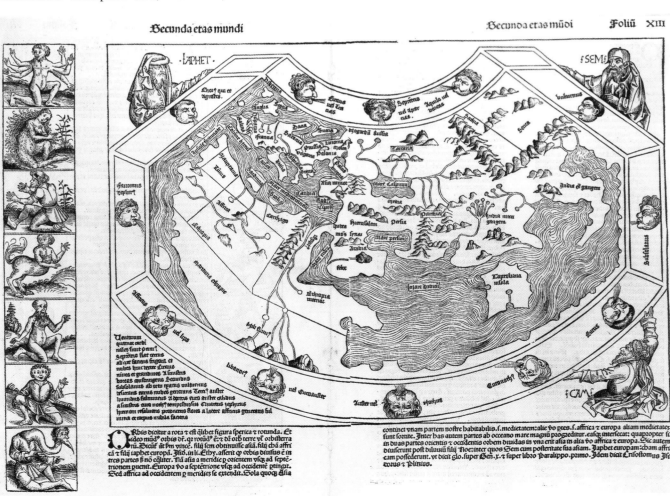

Ptolemy's *Geographia*

The list below of the more important editions of the *Geographia* indicates the importance of both Italy and Germany in the early dissemination of cartography.

1477 First edition with maps published at Bologna.

1478 Second edition; Rome – further issues 1490…

1507 Includes 'new' maps of northern Europe, Spain, France, Poland, Italy and the Holy Land…

1508 In addition to the new maps of the previous edition includes the important World map by Johann Ruysh (see page 36).

1482 Third 'Berlinghieri' text edition; Florence – first issue of Ptolemy to include 'new' maps, i.e., France, Italy, Spain and Palestine.

1482 Ulm edition; first edition published in Germany, first with woodblock maps – all previous being copperplate. Includes first map of Scandinavia. Reissued in 1486.

1511 Venice; edition by Bernard Sylvanus. Maps printed with letter press in red and black.

1513 Strassburg 'Waldseemüller' edition; Includes important new maps of Americas, Asia and Africa (see page 37); reissued in 1520.

1522 Strassburg; re-cut reduced versions of 1513 edition. New maps of the world by Frisius, S.E. Asia and Eastern Asia. Reissued in 1525, 1535 (Lyons) and 1541 (Vienne, Dauphine).

1540 Basle – 'Munster' edition; maps by Munster, published by Henricus Petri. Includes new maps of the world

and the continents and parts of Europe added to the Ptolemaic originals. Reissued in 1541, 1542, 1545, 1551, 1552.

1548 Venice – 'Gastaldi' edition; the first 'pocket-size' atlas includes new maps of sections of the New World.

1561 Venice – 'Valgrisi' edition; enlarged versions of the 1548 edition. New World map added – the first in double hemisphere format. Reissued in 1562; 1564 and 1574 by Ziletti; 1598/9 by Sessa (plates reworked with addition of some ships, monsters, animals etc).

1578 Duisburg – 'Mercator' edition; Gerard Mercator's version – finely-engraved copperplates reissued with plates reworked, i.e., 1584 Cologne, 1605, 1618/19, 1695, 1698, 1704, 1730 at various Low Country locations.

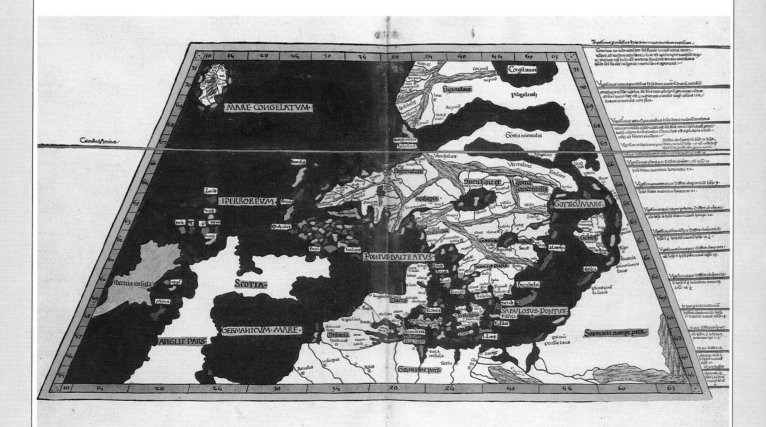

This extraordinary and very rare map is the first printed representation of northern Europe and the north Atlantic and is amongst the four 'new' maps added by mapmaker Dominus Nicolaus Germanus to Leinhart Holle's edition of the 'Geographia' published at Ulm. As such this was the first edition with woodblock maps and the first published north of the Alps.

Untitled

The first printed map of Scandinavia.
Ulm 1482
57 x 33cm 22.5 x 13in

Based on earlier manuscripts, the map includes Iceland and, in this particular example, shows the bright original colours

which are so distinctive of this edition. Maps from the second edition of 1486 may, generally, be distinguished in most instances by a lack of decorative border to the panel of text on the reverse of the map, and, when coloured, by the use of less strident colours – particularly on the sea, where an olivey-green colour as opposed to blue is used.

The first great tracks

In 1487 Bartholomew Diaz rounded the Cape and opened the route to the East. For many centuries European knowledge of the vast riches, precious stones and metals, herbs and spices of the Orient had inspired travellers' tales. Marco Polo's thirteenth-century journey is well-known and his reports often found their way into map detail and illustration (see page 120). As early as the fourth century, the Romans had 'mapped' their world extending well into central and south-eastern Asia.

The finding of a sea route, however, was the key to developing commercial and national ideals, because of the obvious advantages in terms of speed and ease of transportation of men, arms and goods.

The earliest map to break the Ptolemaic tradition of a land-locked Indian Ocean was engraved in Italy by 1493, at the same time as Schedel's map from the *Chronicle*. Francesco Rosselli's map, known in only one copy, shows a completed African coastline, the Indian Ocean and a Ptolemaic conjectural South and East Asian coastline heavily surrounded by islands.

It was, of course, the search for an alternative route westwards to the Indies that inspired Cristoforo Colombo, (Christopher Columbus), to set sail. Born in Genoa in 1451, he is believed to have spent his early working life as a Mediterranean seaman; having worked his way up the decks and through the sea trade, in 1476 he found himself in Lisbon where he became absorbed in the work of Europe's leading cartographers, astronomers and navigators. As a consequence of Marco Polo's reports, and of theories and calculations concerning the size of the Earth and the width of a degree, Columbus calculated that Japan and China lay but a mere two and a half thousand miles across the Atlantic. These theories of the proximity of land to the west were supported by sightings by sailors to the Azores of land vegetation and worked timber on the waves after severe westerly gales. The eastwards elongation of Asia was popularly accepted and at a time when it could not be disproved, why not?

However, Columbus could not find official Portuguese approval and backing for his intended voyage. King Joao II was actively supporting the voyages of discovery along the African coast, which in 1487 were to bring Diaz his success, and Columbus was seen as a brash young boaster whose calculations were dubious and whose demands for financial backing were excessively ambitious.

In 1485, spurned by the Portuguese powers, Columbus who had married into a Portuguese noble family in 1479, left for Spain with his son Diego. His wife had died while still in Lisbon. Columbus, taking the Spanish form of his name, 'Colon', through a mixture of good fortune and determination, was able to put his plans before the Spanish Court. However, it was not until 1491, after further refusals, that with the help of supporting court attendants, he persuaded the Spanish Treasurer, Luis de Santangel, to finance, and Queen Isabella to support his voyage. On the 3rd of August 1492, Colon left Palos, a small port between Cadiz and the Portuguese Algarve, and headed west beyond Cape St Vincent. The Italian-born sailor who had become an accomplished seaman through his Portuguese associations was now sailing from Spain to a land he expected to be Asia but which would be known, after his death in 1506, as America.

On the 12th October 1492, Admiral Colon made his celebrated landfall. His discoveries do not appear on any printed map until 1506 and 1507 when the maps of Giovanni Contarini and Johann Ruysch were engraved. The first of these maps is known in only one copy but the Ruysch map does appear on the market, albeit very rarely (see page 36).

Columbus's three voyages to the New World between 1492 and 1498 were not the only explorations in the western hemisphere of the fifteenth century. Ignoring the disputed legendary claims of Prince Madoc of Wales, Saint Brendan of Ireland, the Zeno brothers from Venice and others from Scandinavia, Giovanni Caboto, born a year before Colombus, also in Genoa, must be considered.

Cabot, his name Anglicized, shared the belief that India (i.e., The Orient) could be reached across the Atlantic, but via a northerly route. Having failed to find support in Italy he settled in London and later in Bristol, one of the most important trading ports on England's Atlantic coast. In England he encountered much the same apathy but was able to raise just enough finance for one ship and eighteen men to set sail in 1496, with authority from Henry VII, to search for unknown lands. Locating Newfoundland, believing it to be Cathay, he reported the timber and the vast fisheries of the area. However, his second voyage, better supported with five ships, had navigational problems and despite 'land-hopping' from Greenland to Labrador, Cape Cod and ultimately the Chesapeake before returning home, was by and large unsuccessful.

The significance of both Columbus's and Cabot's discoveries was simply not appreciated in Europe, whereas the voyages of Vasco da Gama, which took place at exactly the same time as Cabot's, were. Da Gama's voyage, commenced in 1497, completed Diaz's earlier rounding of the Cape. By 1499 da Gama had reached India and set up the first Portuguese trading post at the port of Calicut. Three years later, having returned to Europe in the meantime, da Gama's fleet of 19 ships re-established the settlement, whose original occupants had been massacred, and lay the foundations for the mighty Portuguese trading empire in India.

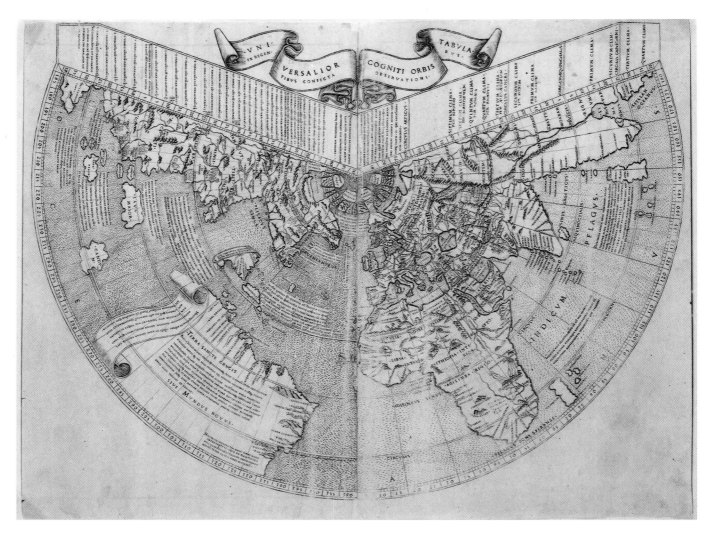

Above
Universalior Cogniti Orbis Tabula ...

Johann Ruysch's important map showing
the New World.
Rome 1507
53.5 x 40.5cm 21 x 16in

*The earliest map of the World, available to collectors, to show
the new discoveries of the Portuguese in Africa, Asia and the
New World. This very rare map was issued separately and
occasionally, bound into some copies of the 1508 edition of
Ptolemy's 'Geographia'. The eastern African coast is well plotted
and the west coast of India shows good detail. Mainland Asia
merges into Greenland and Newfoundland, thus leaving no room
for the fabled island of Sipangu (Japan) which is deemed, in a
panel on the map, to be Spagnola.*

The final stage in this period of European expansion
across the World must be Ferdinand Magellan's
circumnavigation. As with Columbus and Cabot, the
outcome of Magellan's voyage was not his original
intention. Setting out to find a westward passage to the
Indies around South America, Magellan arrived via

Tierra del Fuego at the Pacific Island of Guam in 1521.
Eight weeks later he was killed in the Philippines by
natives, without being aware that he had already visited
Guam when he had sailed eastwards to the Orient years
earlier. Part of his crew subsequently returned to
Portugal and thus completed his journey.

It was still many years before the World map
developed a recognizable shape but the foundations
were laid whereby, within 30 years, the existence of
previously uncharted oceans – the Atlantic, Indian and
Pacific – and, particularly, some of their coastlines,
could be identified.

As these amazing feats of human endeavour and
endurance were taking place so these events were being
recorded and made available to an ever more receptive
public through the new mass communication technique
printing. The coincidence of the great age of discovery
with this new industry is indeed fortuitous for anyone
who appreciates maps today.

The New World maps

The production of Ptolemaic atlases continued
throughout the first half of the sixteenth century. To
the basic corpus of 27 ancient geography maps, different

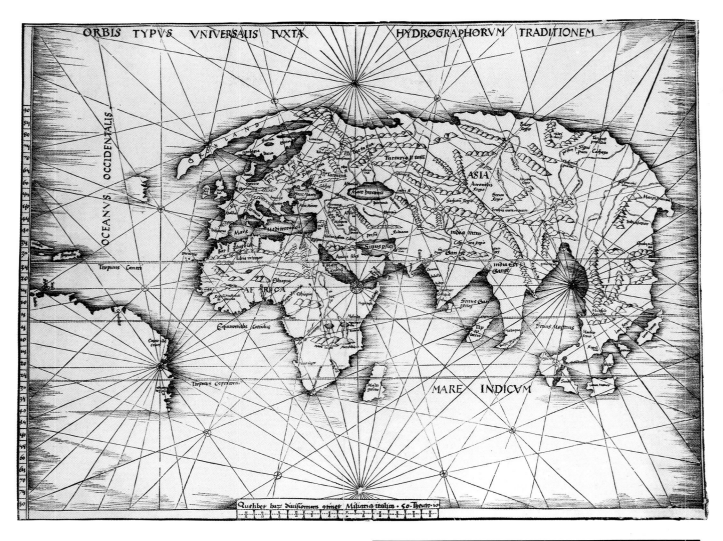

ORBIS TYPVS VNIVERSALIS IVXTA HYDROGRAPHORVM TRADITIONEM

editors or publishers added new, modern maps so that by about 1550 the 'new' maps had become the most important part of the atlas whereas the original ancient maps satisfied a historical or classical interest.

In 1507 Martin Waldseemüller had issued a twelve-sheet World map. This was the best map of its day and was the first on which the word 'America' was used, Waldseemüller identifying that coastline of South America explored by Amerigo Vespucci, whose voyage journal he reprinted. Waldseemüller is best known among map collectors today for the series of 20 'modern' maps which he produced for publication, in Strassburg, for his edition of Ptolemy's *Geographia*. Amongst these were several maps of particular importance: the first map of a specific part of the New World (see page 139), the first detailed maps of Western and Southern Africa (see pages 111 and 112) and the first map to concentrate on Southern Asia.

Waldseemüller's atlas, first published in 1513, was reprinted in 1520. Later, in 1522, 1525, 1535 and 1541 the woodblock maps were reissued having been recut to a reduced size.

Atlas publication in the mid-sixteenth-century was still confined largely to Italy and Central Europe and

Above
The Admiral's Map

From Martin Waldseemüller's important
edition of the *Geographia*.
Strassburg 1513
58 x 44.5cm 23 x 17.5in

This strange woodblock map shows, on the one hand, bizarre cartography (see Greenland and Southern Asia), on the other, the newly discovered islands of Isabella (Cuba) and Spagnola (Santo Domingo). These islands were, of course, those discovered by Columbus some twenty years earlier and a loose reference within the book's text referring to a certain admiral of the Spanish King Ferdinand has been taken to mean Columbus himself. Further detail of the north coast of South America results from the voyages of Cabral.

The map appears in a second edition of 1520 and is accompanied, in the atlas, by a better and more important map of the new discoveries in America (see page 139).

three publications must be mentioned before the appearance of the Low Countries as the leading map-making area.

The *Cosmographia* of Sebastian Munster must rank as the greatest geographical compendium of the period

– an immensely detailed work illustrated with woodcut portraits, scenes, town plans and panoramas, and maps. Born in 1488, Munster was a Franciscan who became Professor of Hebrew at Heidelberg and later at Basle, where he taught Hebrew and, amongst other works, published the first German translation of the Bible from Hebrew. In 1540 his edition of Ptolemy's *Geographia* was published, followed in 1544 by the *Cosmographia Universalis*. Together these ran to over 35 editions published mostly in Basle in Latin, German, French and Italian versions. Munster's particular cartographic importance lies in the number of 'new' maps he introduced and, above all, in the innovative, separate mapping of each of the four continents. The map of the Americas is not only the first map to show the Western Hemisphere separately, but is also the first to show North and South America joined together (see page 135).

During the sixteenth-century, Italian map engraving techniques progressed from the crude appearance of the woodblock to the more refined, elegant lines of the copperplate. In Venice in 1548, Giacomo Gastaldi engraved a fine series of maps for what is seen as the first 'pocket' or miniature atlas. Although the book is

yet another *Ptolomeo … Geographia …* 34 of the 60 maps are modern – most taken from Waldseemüller – but of these there are five new maps of parts of the New World and its islands. Girolamo Ruscelli re-engraved and republished the maps from this atlas in larger format in 1561, and it is this edition of the atlas that includes the earliest double hemisphere World map to appear in atlas format.

Italian 'atlas' production of this period was very

Below
Typus Universalis

The first state of Sebastian Munster's
map.
Basle 1540
36 x 24.5cm 14.5 x 9.5in

A charming, famous woodblock map which in this form and with a second state produced ten years later, appeared in all editions of Munster's 'Geographia' and 'Cosmographia' until 1578. The second state of the map is identified by the initials D.K., for David Kandel, the engraver, in the lower left corner.

Zipagri – Japan – is shown in the East Pacific. The North American mainland is almost split by the fabled sea of Verrazano.

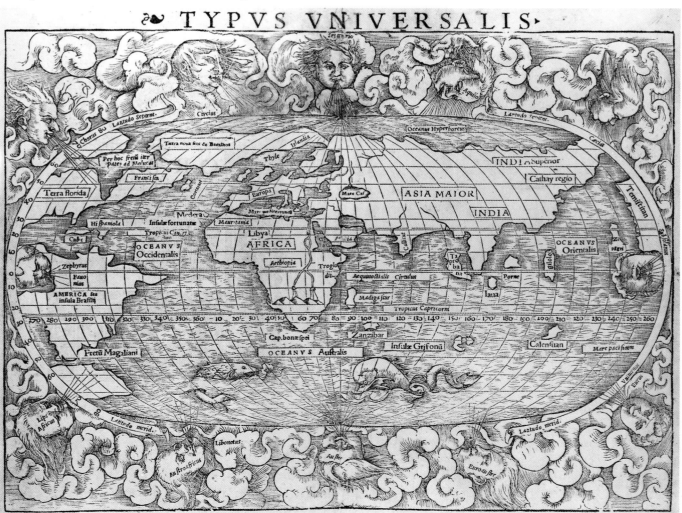

limited, whereas map production was not. Maps of most parts of the World were engraved by a variety of craftsmen and frequently these individual sheets would be bound together in a 'composite' collection, often to a client's specific requirements. For map collections of this type an engraved title page was ultimately designed and these 'atlases' are now termed *Lafreri* atlases after the map seller, Antonio Lafreri (see page 107). However, this is misleading since he was by no means solely responsible for these volumes. During this period many fine and important maps were produced, but because the majority remained unbound they rarely find their way onto the market.

The Low Countries map makers

As the sixteenth century progressed through its third quarter so the Low Countries became the ascending stars in the map making system. Firstly Antwerp, then Amsterdam became the leading map producing centres until the start of the eighteenth century.

Prior to 1570, Gerard Mercator, Abraham Ortelius and Gerard de Jode had each been issuing maps or globes. However, in 1570 Abraham Ortelius issued the first edition of his *Theatrum Orbis Terrarum*, accepted as the first 'atlas' of the World, i.e., the first, bound collection of uniform size and presentation of all known parts of the World. (N.B. The word 'atlas' at this time did not have its present-day meaning – it was Mercator himself in 1585 who first used it in the modern sense (see page 40).)

Ortelius's book of maps became the most successful cartographic production to date, being issued for over forty years in as many editions with text in seven different languages. The first edition comprised some 53 map plates. There were 119 by 1598, the year of his death, and 128 in 1612, after Vrints had taken over publication. Besides the composition of the maps, drawn by Ortelius himself before being engraved by Frans Hogenberg and his assistants for the *Theatrum...*, two other features combine to make it different from previous map books. Each map had, on the verso, a printed description of the area, and Ortelius credited the authors of the original maps, which he himself had copied.

The *Theatrum...* enjoyed enormous popularity and completely eclipsed the rival publications of 1578 and 1593 by Gerard de Jode and his son Cornelis. Ortelius and Mercator might appear now to have been rival map publishers but this was not the case. The two worked closely together and maintained a respect and friendship which was of mutual benefit. Ortelius was the more commercially minded and businesslike of the two, adapting other map makers' work for his publications; Mercator's interest was a more systematic and scientific refining of cartography. It has been

suggested that Ortelius used Mercator's ideas in the concept of the *Theatrum ...* . If so, Mercator, sixteen years his senior, showed no displeasure, only praise. Based in Antwerp, one of Europe's great commercial centres, Ortelius was able to show his friend all the latest maps as they came to him, and equally, Mercator reciprocated by giving advice and encouragement. Both cartographers owed a great deal to the classics – Ortelius's earliest separately published maps included 'historical' ones (see page 40); Mercator's first atlas publication was of Ptolemy's *Geographia*.

In 1585 Mercator's 'Atlas' Part One was issued, comprising 51 maps of Holland, Belgium and Germany. This publication was the first to be called 'Atlas'. The subsequent history of the parts of Mercator's atlas is complex and involves not only his sons and grandsons but also many of the more important names in the Dutch map making industry.

Part Two of the 'Atlas', comprising maps of Italy, Greece and Central Europe, appeared in 1589, and Part Three comprising The World, Continents, British Isles and Northern Europe appeared in 1595. From then on the three parts were issued in one or sometimes two volumes.

Gerard Mercator died in 1594 and his son Rumold, who, himself, died only five years later continued to publish the 'Atlas'. The earlier editions of the 'Atlas' had been published in Duisburg, where Mercator spent most of his life. He had three sons, Arnold, Rumold and Bartholomeus. Only Rumold outlived him, adding the World maps and Europe to the last part of the 'Atlas'.

In 1606 Jodocus Hondius, who had bought the map copperplates from the Mercator family, reissued the 'Atlas' from Amsterdam adding new maps particularly of Africa, Asia and America. Jodocus died in 1612 and his widow and sons, Jodocus II and Henricus continued publication until 1623 when Henricus, alone, persevered. After his brother's death, Henricus, in about 1629 sold a large number of the plates to Willem Janszoon Blaeu but continued the life of the 'Atlas' in conjunction with Jan Jansson who added new plates and who in 1638 revitalized the sequence of publications as the *Atlas Novus*. Jansson's efforts, in direct competition with the Blaeu Family, produced, within ten years a nine-volume Dutch edition in German and Latin. In itself, this was a magnificent work and would, at any other period, have been the highlight of a nation's map production but for the work of the Blaeu family.

Before we proceed to the Blaeus' work it is necessary to back-track in time to the end of the sixteenth century. We have seen the innovation of Ortelius's 'Atlas' and the development of Mercator's. At the same time two other publications on the European

Ortelius's *Theatrum*

Abraham Ortelius, born in 1527, is not only famed today as the compiler of the first 'atlas' but in his own day was highly regarded as a map collector and authority on historical cartography. His interests in history and geography and his trade in coins, antiquities, books and maps led him to be, during the last thirty-five years or so of his life, one of Europe's foremost cartographers.

Although from an early age he and his sisters, Anne and Elizabeth, had worked as 'Kaartenafzetters' – map colourists – Ortelius did not actually produce any maps of his own until the mid-1560s by which time, in his mid-thirties and unmarried, he had travelled extensively, trading in Italy and Germany and had also worked in conjunction with some of the leading Antwerp merchants. In these various roles Ortelius had made connections with numerous other historians, topographers, cartographers and others engaged in the map making business and he was able to call on their assistance in compiling his *Theatrum Orbis Terrarum*. His friendships with Gerard Mercator and William Camden in England are well known, Ortelius, like many other like-minded reformists having taken refuge in London during the religious persecutions of the 1560s and 1570s.

The compilation of the fifty-three map sheets in the first edition of 1570 must have taken some years and there is dispute as to whether Mercator, Ortelius himself or associates of Ortelius was or were, responsible for the initiation of the project. However, there is no doubting the success of the volume which, in 1573, was expanded by a further 16 maps.

By the time of Ortelius's death in 1598, the atlas could be made up of some 119 map sheets (of which a large number were replacements for earlier plates) and a further 36 historical, or classical maps. Until that time, numerous editions in Latin, others in German and Spanish, had been issued. In 1602 further editions in German and Spanish were issued by Jan Baptist Vrients who had added new maps. In 1606 the same publisher produced the maps for the only English text and most complete edition of the *Theatrum*. This volume had 123 modern maps and 38 classical plates and had the text printed in London on the verso of the already printed up sheets which had been done in Antwerp. As such, this is one of the rarest and most interesting editions. Finally, editions in 1612 in Latin, Spanish and Italian, brought to a close the publication of this great atlas.

Later to be used at reduced scale in his 'Parergon' (the classical atlas added to the 'Theatrum') this finely-designed map reflects Ortelius's abiding interest in the classics. Ortelius himself takes credit as engraver of the map while the signature of Hogenberg implies his hand in the engraving of the portraits and more decorative elements of the piece.

The medallion portraits are of Romulus

Romani Imperii Imago

A very rare separately-issued map by Ortelius.
Brussels 1571
70 x 49.5cm 27.5 x 19.5in

and Remus – the genealogical 'tree' describes the lineage of the Roman emperors while the strapwork bordered panels explain

the history of the Empire. The dedication between the title and the portrait is to Francisco Usodimaro, a patrician of Genoa and lover of the Classics. Only a very few of Ortelius's separately-issued maps have survived and this is interesting in having, at lower left, the manuscript note 'Aestimato 8 Styvers' being the price originally asked for this mapsheet (a styver was a unit of currency used in Flanders).

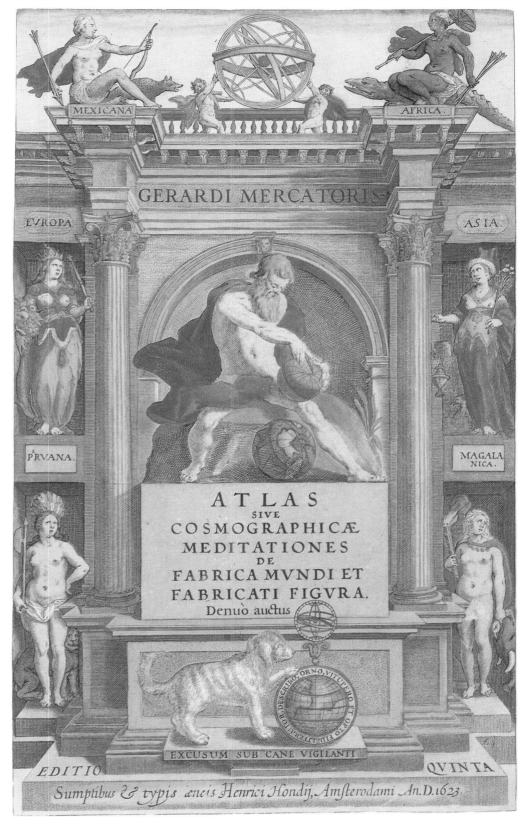

Left
Atlas

Title page from the Henricus Hondius edition of the compilation of maps begun by Gerard Mercator. Amsterdam 1623
24 x 39.5cm 9.5 x 15.5in

Female figures representing the continents – Europe, Asia and Africa, and Mexico, Peru and Magalanica for the Americas, surround a Herculean figure. The name 'Atlas' was chosen after the Moroccan King (hence, Atlas Mountains), renowned as a student of geography and astronomy.

mainland were breaking new ground. These were Braun and Hogenberg's book of town plans and Waghenaer's chart book of the European Atlantic coasts. Also, in England, Christopher Saxton had produced the first National Atlas of the British and Welsh counties.

Each of these three atlases will be treated in more detail in their respective sections (see pages 65, 59 and 79, respectively) but no resumé of the cartographic development of this period would be complete without their mention and due consideration.

The Blaeus

The work of Willem Janszoon Blaeu and his son Joan epitomize the 'golden age of Dutch cartography'. Their

Below
Orbis Terrae Compendiosa Descriptio

Rumold Mercator's copy of his father's great map.
Geneva, later Duisburg and Amsterdam 1587
52 x 29cm 20.5 x 11.5in

Gerard Mercator's great wall map of the world, issued in 1589, was reduced and issued by his son. First issued in Strabo's 'Geography', copies of the map more generally found are usually of the Atlas versions published from 1595 onwards. Numerous different versions of Gerard's original, and this version, were published – testimony to their influence and popularity.

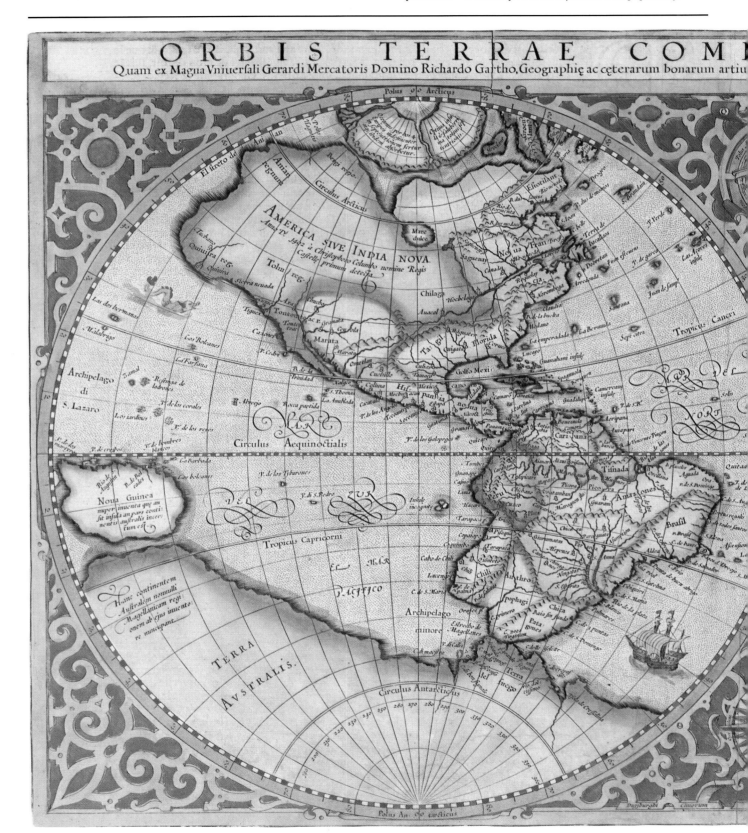

output throughout the seventeenth century reflected the finest features of Dutch map making at a time when the Dutch Empire, with Amsterdam at its heart, was at the height of its wealth and power. Willem Janszoon was born in 1571 and spent his early working years studying to be a clerk in the family herring-trade in Amsterdam. His dislike of this led him to study

mathematics and, in 1594, to become a student of the celebrated astronomer, Tycho Brahe.

Brahe was one of the foremost astronomers of the period and had been granted permission from the King of Denmark to set up a school and observatory on the island of Huen (see page 171), where many of Europe's leading astronomers and cosmographers were to receive

training. After a relatively short period Blaeu, who had greatly impressed his tutor, returned to Amsterdam and set up in business as, initially, a globe and instrument maker, and very soon as a publisher and chart and map maker.

From 1596 Blaeu's major output was of terrestrial and celestial globes, issued in pairs in various sizes and of finely made navigating and astronomical instruments. In keeping with his academic training he was an active astronomer himself and is credited with the identification, in 1600 and 1604, of two previously unrecorded stars. In 1604 the publication of Blaeu's first map, of the Seventeen Provinces, is recorded and this was followed by other single sheet maps of various European countries. Of more importance though were the wall maps of the World and four continents, issued from 1605. The World map, on twenty sheets, totalling some eight feet across, was outstanding in its design and, despite some inaccuracies, was the best of its period. It remained the prototype World map used by many other map makers until superseded by his son's in 1648. Unfortunately, only one example of this great map is known to survive and that is in very poor condition.

Blaeu's next major publication was the atlas of European coast charts, *Het Licht dee Zeevaert*, first published in 1608. It appears that Willem Janszoon intended to produce atlases covering the whole earth since this is advertised as the first volume of four, the final volume to include coasts of all non-European countries – these, however, were never published. The European volumes were reissued many times and, in 1623, replaced by *Der Zeespiegel*, an updated and revised book containing over one hundred charts.

In 1630 Willem Janszoon published the first land atlas. After thirty years of separately-issued map publication during which time he had also published books on a great variety of subjects including the sciences, classics and literature, the time was right, in commercial terms, for a new atlas.

As we saw earlier the Mercator/Hondius atlas, which replaced that of Ortelius, was the most popular early seventeenth-century atlas on the market. In 1629 Blaeu had purchased about 40 of the Jodocus Hondius II map plates and these formed the bulk of Blaeu's *Atlantis Appendix*, which also included many of those single sheet maps issued by Blaeu over the previous fifteen or twenty years. It is interesting to note that by this time the business rivalry between Willem Janszoon, working in conjunction with his son Joan (Joannes), and the neighbouring map publishing firm of Joannes Jansonnius, coupled with the similarity in their names made it necessary for W J to incorporate the family nickname, Blaeu, into his business name.

Joan Blaeu worked with his father enlarging the *Appendix* in 1631, and then publishing its replacement

Left
Gerardus Mercator ...
Iudocus Hondius ...

Two of the greatest figures in the development of map making and atlas production.
Amsterdam *c*1620
44.5 x 38cm 17.5 x 15in

A fine posthumous double portrait which was engraved for inclusion in the Mercator/ Hondius series of atlases from 1619. The two men are surrounded by globes, maps, atlases and navigating and cartographic instruments.

entitled either *Novus Atlas* or *Theatrum Orbis Terrarum* from 1634 or 1635. From a little over 200 maps in two volumes, this publication expanded to around 400 in six volumes by 1655. Willem had died in 1638 and his other remaining son, Cornelis, had joined Joan in business. However, Cornelis died before 1645 and Joan continued his father's business to produce the famous *Atlas Major*.

Although successful, the large atlas which Jan Jansson had been putting together was not such a complete work as Joan Blaeu was intending. Not only was the production rushed but the actual maps and assembled volumes were of a rather 'ad hoc' nature – in most cases they were simply reissues of earlier volumes. Although some of Jansson's maps may be preferred on a visual design basis (a personal and subjective valuation), Blaeu's publication was undoubtedly better printed, on better quality paper, with better hand colour, and available in superior bindings – overall a more elegant and impressive publication. The various editions of the *Atlas Major*, published between 1662 and 1672 vary from nine to twelve volumes, containing in all around 600 maps, in addition to which there is an interesting series of plans and diagrams of astronomical instruments and the Observatory at Huen (see page 171).

In 1672, a fire at the warehouse destroyed much of the Blaeu's stock of copperplates and printed material and a year later Joan died. Remaining stocks were auctioned over the next few years and so their life prolonged into the next century.

In the wake of the *Atlas Major* by Jansson and Blaeu two other publications must be mentioned; the magnificent celestial atlas of Andreas Cellarius, published by Jansson in 1660 and the equally fine sea atlas of Pieter Goos of 1666. Each of these are discussed in their relevant chapter (see pages 173 and 61 respectively).

Map making in France and England

Around 1700, although the Dutch map making industry remained active, there is an identifiable shift of importance of production in Europe. From about 1650, particularly through the efforts of Nicolas Sanson, French map making, especially of North American regions, became more important. Maps by Sanson and du Val did not compete in any way with the Dutch productions in terms of artistry but were often more up-to-date. However, in 1681 Alexis Hubert Jaillot published in Paris and subsequently in Amsterdam from 1696, a superb series of finely-engraved, large-scale maps of all parts of the world. These maps were enlargements of Sanson's, were as decorative as the comparable Dutch maps and were twice the size. It was an Amsterdam publisher, Pierre

Mortier, who took over publication of the Sanson/Jaillot atlas and whose son, in conjunction with his partner Jean Covens from 1730, issued the work of another great French map maker, Guillaume de L'Isle.

De L'Isle's influence can be traced throughout the eighteenth-century in the many reprints of his work or in the great number of maps which give credit to de L'Isle's originals. Of even greater importance is Jean Baptiste d'Anville who, around 1750, produced detailed large-scaled maps, many of which found their way into English publications.

Excluding John Speed's *Prospect* of 1627 (see page 46), the English did not contribute much to the development of European cartography until the mid-eighteenth century and though from around 1670 the likes of John Seller, Robert Morden, John Senex and Herman Moll had produced good maps, there was little in the way of cartographical innovation. However, by 1750, Britain's developing role in the World and the need for detailed mapping of the areas within its sphere of interest, produced a new brand of English map maker. Following on from the traditions of the English county atlas, the home demand was now for large-scale county maps – the forerunners of today's *Ordnance Survey*.

In India, America and the West Indies, the work of English surveyors and map makers was the best available and on account of Captain James Cook, England was able to answer many of the outstanding questions regarding the great Pacific Ocean. When we look at the mapping of individual areas we shall see how important the English map makers of the period from around 1750 were. John Harrison's invention of the marine chronometer, (a maritime clock that could keep accurate time and from which a ship's longitude could then be calculated), was taken on trial by Captain Cook and subsequently facilitated navigation and revolutionized marine surveying techniques.

By 1800 map making was moving into a new phase, i.e., more precise requirements, more accurate instruments and finer techniques encouraged cartographic science rather than cartographic art.

John Speed's *Prospect*

John Speed is as well known now for his maps of the world and its parts as he is for the famous series of county maps which are so distinctive. His county atlas, *Theatre of the Empire of Great Britain*, was first published in 1611 and had come about through Speed's requirement to produce an atlas volume to accompany his history of Great Britain which had been published some six years earlier.

Following the lead of Dutch map sellers, Speed produced his World Atlas of some 21 maps which was published, two years before his death, in 1627. This was the first World Atlas produced and published by an Englishman in England and maintained the decorative and informative qualities of his county maps. The reverse of each map had a printed letterpress description of the country shown, and, in the continental style, excepting those of the world, Greece and Bermuda, had panelled borders all round showing inhabitants of that area and their major cities. Speed's atlas is, therefore, the only one ever published in which the design of the maps is predominantly 'carte à figures' (see page 27).

The maps in the first edition are as follows: World, Asia, Africa, Europe, America, Greece, Roman Empire, Germany, Bohemia, France, XVIII Provinces of the Low Country, Spain, Italy, Hungary, Denmark, Poland, Persia, the Turkish Empire, China, Tartary and the Bermudas. In addition, in 1676, Thomas Bassett and Richard Chiswell had included maps of Virginia; Maryland; New England and New York; Carolina; Jamaica and Barbados; East India, Russia; Canaan; and the 'Invasions of England and Ireland'. Most of these latter maps were engraved by Francis Lamb and are known as 'Speed maps' despite his death of many years earlier in 1629 and despite the particular maps' lack of originality being based on earlier standard material already previously published.

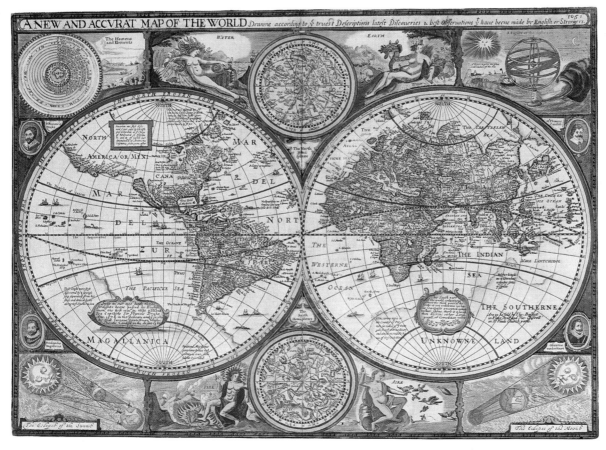

This fine map, although lacking the more delicate artistry of many comparable Dutch engravings of the period, is one of the best-known and most sought-after of all World maps. Besides being a particularly decorative item – with English, as opposed to Dutch or Latin titles, the map is the first obtainable World map to show California as an island. An earlier map, by William Grent, 1625, is known in only a handful of copies but may have provided some of the cartography for this engraving.

Symbolic representations, typically,

A New and Accurat Map of the World

John Speed's famous, decorative double-hemisphere map.
London 1627-1651-1676
52.5 x 39.5cm 20.5 x 15.5in

illustrate the four elements – Earth, Air, Fire and Water; diagrams of an armillary sphere, eclipses, celestial charts and the planetary system surround the map, along with miniature portraits of Drake, Cav-

endish, Van der Noort and Magellan.

Cartography of the US coast shows evidence of recent reports, Le Maire's 'Straight' was only recently shown, and Speed shows some doubt over the very incomplete coastline for the southern continent. First published in 1627 the example of the map shown here has the added date 1651 in the top right corner, and, in the Southern Continent region, the imprint of the publishers – Basset and Chiswell – identifying 1676 as the date of publication of this particular edition.

Blaeu's *Atlas Major*

By any standards the map making and publishing achievements of the Blaeu family during the first three quarters of the seventeenth century stand favourable comparison against the production of other periods. The output of globes, sea charts and land maps reached its most memorable peak with the publication, in 1662, of the first edition of the *Atlas Major* or, in Latin, *Atlas Maior*.

Dutch map production during the seventeenth century has acquired the title of the 'golden age' of cartography and Blaeu's *Atlas Major* epitomizes the style and quality of the period.

Publication of the atlas, whose volumes were based on, and extensions of, Blaeu's earlier *Theatrum* was as follows: *Atlas Major* Latin text, 12 volumes, 1662; *Le Grand Atlas* French text, 12 volumes, 1663 and 1667; *Grooter Atlas* Dutch text, 9 volumes 1664; *Atlas Major* Spanish text, 10 volumes issued from 1658-1672. A further German issue of 1667 is recorded although it

appears to be a composite as opposed to an intended complete publication. Each edition comprised around 600 maps excepting the Spanish whose final volume – of maps of Africa and America – was lost in the fire which destroyed much of Blaeu's stock at their warehouse in 1672.

Joannes Janssonius's eleven-volume publication of his *Novus Atlas Absolutissimus* in 1662 would have spurred the rival Blaeu family in producing their great atlas and, although many of the maps in Joan Blaeu's atlas have their origin forty years or more before this publication the *Atlas Major* proved far more popular: the quality of engraving and typography, of paper used, of binding offered and of 'in-house' colouring of the maps produced an atlas of beauty and distinction worthy of presentation to foreign dignatories as a symbol of the Dutch Empire and the expertise of its craftsmen.

The contemporary colouring of Blaeu

maps is distinctive in tone and care of application, and is frequently heightened with gold embellishments. Occasionally, the work of an expert map painter' 'afsetter van caerten' might be required for a special presentation copy. The artist Dirck Jansz van Santen is well known as such a figure.

The *Atlas Major* was, when published, one of the most expensive books ever produced and now good sets attain prices around one hundred thousand pounds when offered in auction. Despite the age of some of the map plates used, others showed the most up-to-date cartographic information available, and the atlas, as a whole, bore favourable comparison with any other publications of the period. However, after the fire and the death of Joan Blaeu a year later, and with the general decline in Dutch map making, in favour of the French, the great Blaeu publishing business was dissolved – leaving, for posterity, a great atlas production.

First issued as a separate publication in 1617 – signed Janssonio before the change of name to 'Blaeu' (c 1621) – this famous 'carte à figures' was the standard map of the New World used in all Blaeu atlases.

Identification of the date of publication of loose maps from a series of atlases in so many editions may prove difficult. Frequently, in the absence of any alterations to the face of the copperplate, it is necessary to examine the verso text, page number and the 'signatures' of text pages (usually one or more letters indicating the sheet's

Americae

One of the four maps of the continents which appeared in the Blaeu's World Atlas.
Amsterdam 1617-c1650
55 x 41cm 21.5 x 16in

location within a volume) against a standard reference. This particular map maintains the original peninsula form of California although, during the lifetime of this

map's publication, many map makers preferred the island theory – even Joan Blaeu in his double hemisphere World map from 1658 showed an island, although in the same publication the America map showed a peninsula.

Panelled borders at each side show full-length portraits of inhabitants of the Americas and the oval views, at the top, include panoramas and plans of major New World cities including Havana, Santo Domingo, Cartagena, Mexico, Cusco, Potosi and Rio de Janeiro.

French Theoretical Cartography

From about the mid-seventeenth century, French map making had achieved a respected international acclaim and the maps of Sanson, de Fer, de L'Isle and others were accordingly being re-printed outside France. However, about 1700, a new 'approach' to map making was being practised in France – known now as 'theoretical' cartography. At the forefront of modern cartography was Guillaume de L'Isle who, from 1700, published maps and atlases of great quality; these were reissued many times after his death in 1726 by Philippe Buache, his nephew Jean Nicolas Buache, and, subsequently, by Dezauche.

Despite publishing the precise, accurate maps of his late father-in-law, Philippe Buache was also responsible for some of the most bizarre and fanciful cartographic outlines ever published. Some of his maps were innovative and showed interesting physical features, while others were pure conjecture. Buache's cartography, as one of the leading 'theoretical' cartography practitioners, produced some fantastic outlines with special emphasis on the north and west coastline of North America and on the Australian and southern continental land masses.

Many French mapmakers of the mid-to late-eighteenth century followed the lead of Buache, in particular, and Dezauche, and it was not until the voyages of Cook, Vancouver and others that their concepts were finally disproved (see page 131).

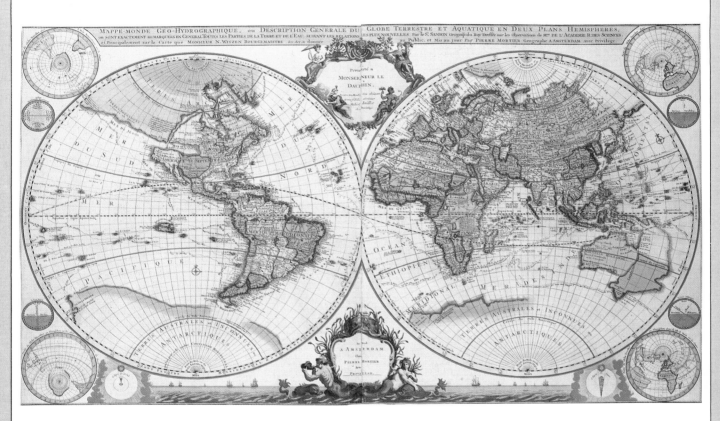

Mappe Monde Geo-Hydrographique ...

Alexis Hubert Jaillot's magnificent,
curious map of the world.
Amsterdam c1700
96.5 x 56.5cm 38 x 22in

This fine, rare map – here in strong original, wash colour – appeared in some issues of Pierre Mortier's edition of Jaillot's enlarged versions of Sanson's maps. As such, it is a cartographical curiosity – coming, apparently, from a background of maps as correct as contemporary information would allow and yet showing outlines which had no basis in fact whatsoever. This map would be seen, therefore, to be one of the earliest examples of French theoretical cartography.

Four features may be easily identified as worthy of note. 1. The possibility of a south polar landmass had long been assumed and the outline here is of no remarkable difference to other maps of the time, but the following features are peculiar to this series. 2. The curve of South America towards the west. Even Guillaume de L'Isle showed this outline in his earliest maps, but he corrected his detail –many others did not. 3. The north-west coastline of North America. Versions of this optimistic theory appeared throughout the century. The 'Mer de l'Ouest' as it became known and the channels accompanying it joining the Pacific with the Hudson Bay derived from the fictitious reports of the Spanish navigator Juan de la Fuca and continued on maps until Cook and Vancouver had sailed and plotted that coastline. 4. The extended Australian eastern seaboard is regularly encountered and, as the earlier misconception, was disproved by Captain James Cook.

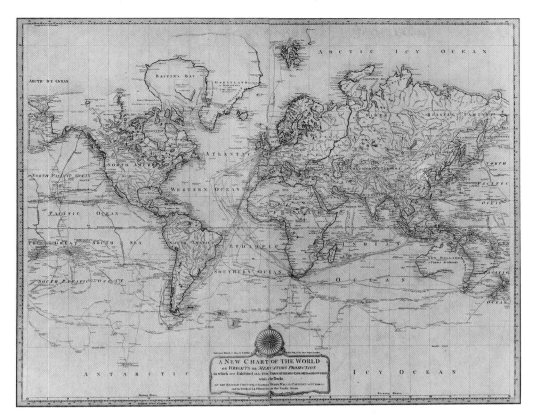

Left
A New Chart of the World

A summary of knowledge
at the start of the nine-
teenth century, by Laurie
and Whittle.
London 1799-1800
92 x 72cm 36 x 28in

*First published in 1799, this
large-scale, detailed map is
evidence of the quality of
English map making and
shows, by the tracks of explorers Byron, Wallis, Carteret,
Cook, La Perouse and others,
the great activity of the preced-
ing twenty and thirty years
which confirmed the outlines
of the continents.*

*The inset, below, of the
Australian region, is in itself
an annotated development of
that country's coastline.*

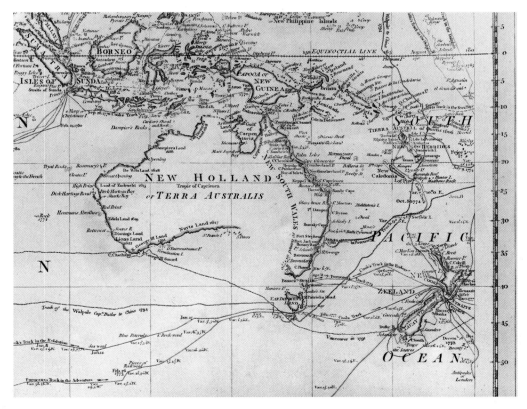

Left

Detail ...

*This section of the right hand
side of the large map above
illustrates the tracks of, in
particular, Cook's voyages of
1770 which defined the shapes
of New Zealand and the
Australian East coast.*

*Note how the map maker
here has held with established
fact and left blank those coastlines of south and north Australia, New Guinea and other
islands, of which he was uncertain.*

*'New Zealand' is named
such though the original native
names for each island are also
used. Australia, though, is not
identified by its modern name.*

WORLD MAPS

For many people, an expensive, decorated double hemisphere map of the World is the first image which enters the mind when the subject of 'antique maps' is mentioned. However, the variety of World maps available to the collector extends far beyond this stereotype and does not necessarily involve great financial outlay.

A comparison of World maps of different periods will show an evolution of European geographical knowledge as well as a constant redefining of political, i.e., national boundaries. Add to this the differing designs and varying qualities of production, and in very few cases do two World maps appear to be identical. Beyond the purely geographical content – the shape of landmasses and locations of places – each World map is conceived on a specific projection. The map maker's

time-old problem has always been how to represent the details of a circular object, i.e., the World, on a flat piece of paper. This could only be achieved with reasonable accuracy on a sphere itself – a globe.

It is this problem which has led map makers, both past and present, to invent a variety of different 'projections' by which a more accurate representation might be obtained. Any of the following projections might be encountered:

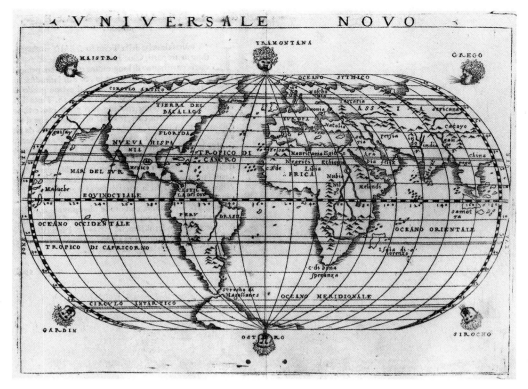

1. *Conical, Gnomonic* Ptolemaic (see page 11).

2. *Planispheric* The World shown completely within a circle (see page 11).

In medieval maps the 'T-O' form (see page 32) evolved into a more detailed map of the known World, invariably with Jerusalem at the centre, but such a map did not extend beyond the bounds of ancient Ptolemaic geography.

Planispheric projections in later periods were invariably associated with specific scientific theories (see page 56).

3. *Cordiform, fan-shaped, heart-shaped* As a derivation and expansion of the traditional, conical projection, map makers from the start of the sixteenth century began using a variety of projections which have assumed the titles suggested by their shapes (see page opposite).

4. *Oval* Deriving from Francesco Rosselli's map of 1508 (known in only three examples) this shape was copied by Bordone (see page 58), Gastaldi, Munster (see page 38), and, most famous of all, Abraham Ortelius. This is the first projection which attempts to equalize distance distortion north and south of the equator, i.e., the first World map format with the equator drawn across the World's centre.

5. *Double hemisphere* This first appeared in 1554 in Italy in a map by Michele Tramezzino later in 1561 and, in an atlas in Venice by Girolamo Ruscelli. The double hemisphere halved the E-W latitudinal inaccuracy of the Oval projection, and on account of its well-balanced design, allowing for the incorporation of decoration or descriptive panels, and its inevitable setting of the New World against the Old, proved over the centuries to be the most enduring form (see the following page).

6. *Polar projection double hemisphere* This is a variation of the principle above which allows a map's emphasis to be on the North and South Pole and makes it possible, for instance, to illustrate a theoretic balance of a terrestial sphere against a maritime sphere. Polar projections were published during the sixteenth century

Orbis Terrarum Typus ....

The first World Map with elaborate, pictorial borders.
Amsterdam 1594
56 x 40cm 22 x 16in

This beautiful map from Linschoten's 'Voyages' combines the skills of two of the most respected map makers and engravers of the day. Petrus Plancius and Jan van Doetichum, whose signature is visible at lower left, worked together on many map productions and this is one of their best known.

This map is the first to use elaborate pictorial borders representing the peoples, animals and environment of foreign parts and established a tradition which was maintained by most Dutch map makers throughout the next century and by numerous others of various nations over the next two hundred years. In each corner are female representations of the four continents: Europe, an elegant crowned figure holding a cornucopia and a sceptre, a helmet, a lute and symbols of wisdom at her feet; Asia, an elaborately robed figure seated on a rhinoceros and holding an incense burner, a casket of baubles at her feet; Africa, an almost naked figure riding a crocodile armed with bow and arrows; America, entitled Mexicana, an Amazonian figure seated on an armadillo.

Between the figures and the celestial spheres at top and bottom are further illustrations of the animals, people and habitations of these exotic places. Much of the illustration here, as in numerous subsequent Dutch map design was inspired by the illustrations in the reports published by de Bry a few years earlier.

This map by Plancius was copied almost line-for-line by others but few of the subsequent seventeenth-century World maps came close to matching this for combination of content and decoration.

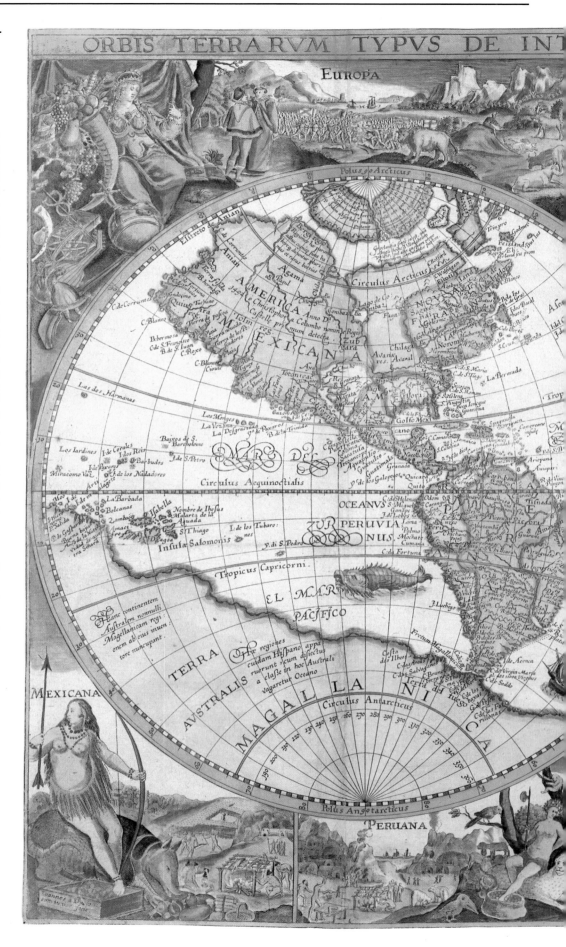

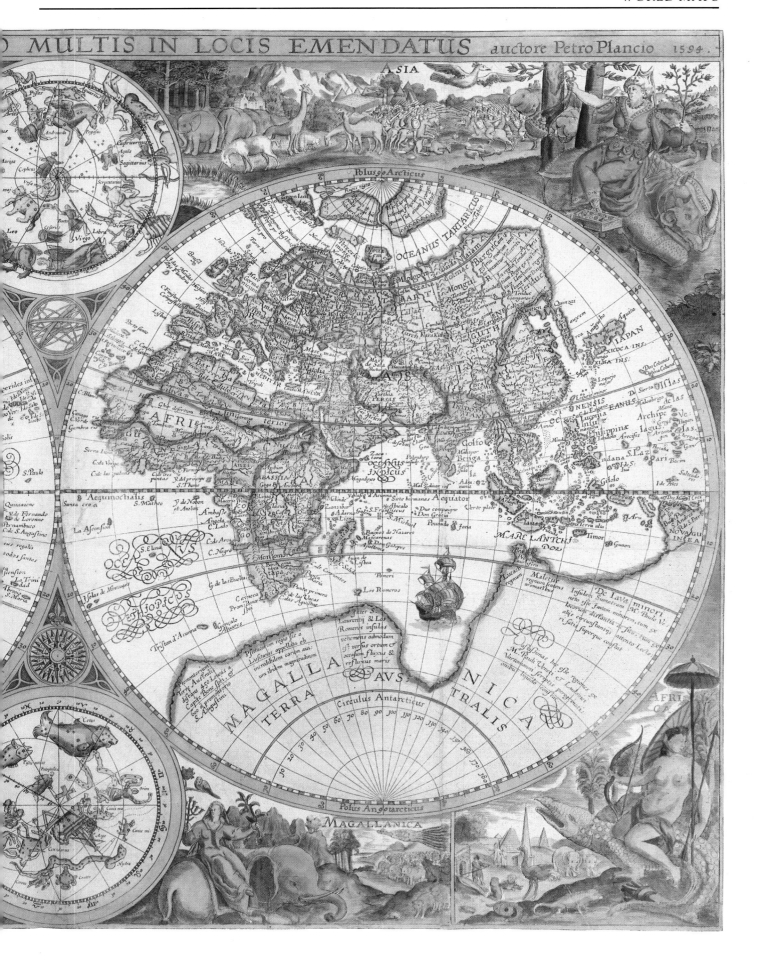

MULTIS IN LOCIS EMENDATUS auctore Petro Plancio 1594.

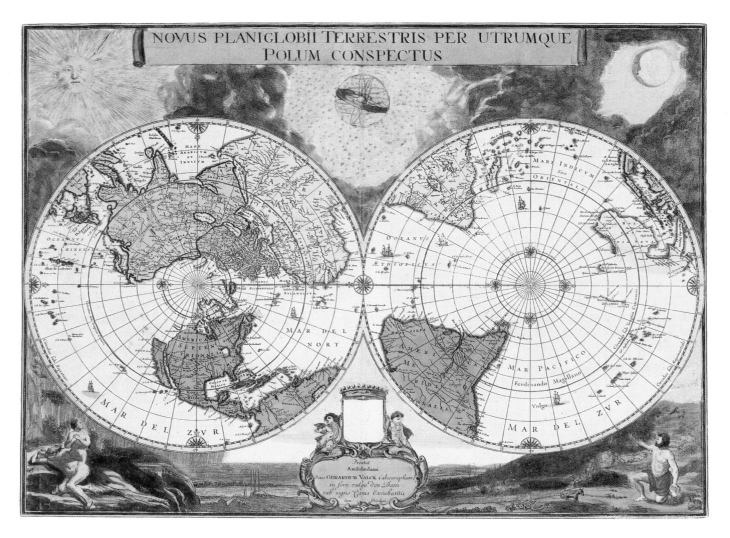

**Novus Planiglobi Terrestris Per Utrumque
Polum Conspectus**

A polar double hemisphere map
published by Gerard Valck.
Amsterdam c1710
54.5 x 41cm 21.5 x 16in

*This very decorative map in strong original colour is surrounded
by scenes of the elements, Earth and water, Adam and Eve,
animals, the Sun, the Moon, the stars and an armillary sphere.
The original copperplate for this map is believed to have been
engraved for Joan Blaeu, probably for incorporation into the
'Atlas Maior', since Blaeu's name can be identified, on some
examples of the map, beneath the imprint of Valck.*

and, in greater numbers, during the seventeenth but
there are relatively few compared with the standard.

7. *Segmented* Practically speaking, this was the most
accurate format but also the most difficult to read unless
the segments or gores were actually stuck onto a globe.
Many of the 'segmented' maps which are now available
to collectors were originally intended to be cut and
pasted onto a globe. The word 'gore' defines such a
segment (see page 18).

8. *Flat or Mercator projection* The most famous and
important innovation in World map projections was
Gerard Mercator's great World map of 1569. By
'opening out' the top and bottom of the globe and
allowing expansion to the same width as the equator,
distances S-N and N-S of the equator were increased;
however, it was now possible for navigators to plot a
compass route of a straight line on the map. Although
it was many years before this new charting method was
fully appreciated and utilized, Mercator's projection
was used throughout the seventeenth and much more
frequently from the eighteenth century onwards.

To define 'important' World maps is no easy task
since there are a large number of criteria upon which
judgement might be based. Furthermore, it is not the

purpose of this book to dwell at any length on
cartographical productions which may not reasonably
be expected to come onto the market. Therefore, I will
list and illustrate briefly only some of the available
maps which are of particular interest or significance.

The importance of any particular map may depend
on a number of factors: its age, the geographical features
shown (or not shown), the author, engraver or

publisher, its rarity or its decorative features.

Around 1500, cultural Europe, in the grip of the High Renaissance witnessed the opening up of New Worlds and the maturing of the earliest form of mass communication – printing.

In 1507 the ancient Ptolemaic cartography mould was broken and although the images remained, map printing and publishing had moved into a new era. In this year Martin Waldseemüller's large, separately-issued Wall map and Johann Ruysch's conical projection map both recorded the discoveries of the New World. A year earlier a map by Contarini and Rosselli had shown this detail but both this map and that of Waldseemüller are known only in one copy: although the Ruysch map was issued separately and is very rare, some copies were bound into 1508 editions of Ptolemy and the map does occasionally come onto the market. It is thus the earliest available map showing New World discoveries (see page 36).

Of much less rarity but of great interest to collectors is the Waldseemüller map of the World (see page 37), which appeared in his 1513 and 1520 editions of *Geographia*. This so-called 'Admiral's Map' does not have as much detail as his earlier wall map and is associated with Columbus on account of a textual passage referring to observations made by the 'Admiral', i.e., Christopher Columbus. The 'Admiral's' map was copied, on a slightly reduced scale, by Laurent Fries in 1522 for his edition of *Geographia* which also included another world map – not dissimilar to the first – but identifying the New World landmass as 'America' (see page 134).

Sebastian Munster, whose *Cosmography* was to be so important, had published his own edition of Ptolemy, in 1540, to which he added modern maps of the World, the Continents and parts of Europe. These maps are not rare but are most sought after on account of their woodblock style, charm and design (see page 38).

Over the next thirty years some beautifully engraved and important maps were printed in both Italy and the Low Countries and it was in Antwerp, that Ortelius's famous atlas was published in 1570. This, of course, included a world map, based on his friend and colleague Mercator's great Wall map of 1569 rather than his own eight-sheet map of 1564.

One of the most curious of World maps was published in 1581 in Heinrich Bunting's *Itinerarium*. This 'clover-leaf' map derives from Medieval Mappa Mundi and has Jerusalem at its centre with the continents of Asia, Africa and Europe as the three leaves – a design apparently inspired by the coat of arms of Bunting's home town, Hanover.

Gerard Mercator's work had not, until 1585, been issued in atlas form. In 1587 his son, Rumold, prepared a single sheet version of his father's map which was issued first in Strabo's *Geographia* and from 1595 in issues of the 'Atlas' (see page 42).

In 1590, Theodore de Bry, in Frankfurt-am-Main, began publication of his reports of voyages and travels concerning the New World, the *Grands Voyages*. In addition to superb maps, this contained some of the earliest explorers' illustrations to emerge from Virginia, Florida and Latin America. Many of these illustrations of natives, animals and plants were used within map design over the next hundred years or so but it was Petrus Plancius in 1594 who first incorporated them into a World map and thereby created a new vogue in map decoration (see page 52).

Until this time any decoration on maps was stylized, taking the form of clouds, strapwork, windheads and, very occasionally, figures or portraits. From now on, vignette illustrations and border decorations were used to add copious information to the basic cartographic content.

Innumerable World maps from Dutch, English and French publishers over the next century or so sought to educate the reader with illustrations of each continent's inhabitants, culture, flora and fauna, or with symbolic representations of the seasons and the elements through actual or mythological pictures.

Map making in England had little foreign interest until the publication, in 1627, of John Speed's *Prospect of the most famous parts of The World*. Apart from the English text edition of Ortelius's atlas issued in London in 1606, this was the first World atlas produced in England. It contained a map of the World, a map of each of the continents, and sixteen others including Bermuda and parts of Europe and Asia. This was the first atlas to promote the theory of California as an island (see page 160), and the World map, with its multitude of detail, though not as finely engraved as comparable Dutch publications, is now particularly sought after by both serious collectors and those who want a single, charming, decorative piece.

With the exception of Abel Tasman's discoveries of the early 1640s, during the seventeenth century relatively little major seaborne exploration was carried out – the European powers were quite busy enough fighting each other at home or establishing and protecting the colonies they had acquired during the previous hundred years. World maps of this period reflect the internal development, particularly in North America and the western hemisphere although, inevitably, they lack the detail of regional maps. As we have seen the century was marked by prolific and magnificent cartographic publications, and it came to a close with the fine work of the Italian, Vincenzo Coronelli, who issued atlases, separate maps and, most importantly, globes.

Coronelli's work is notable by its precision and use

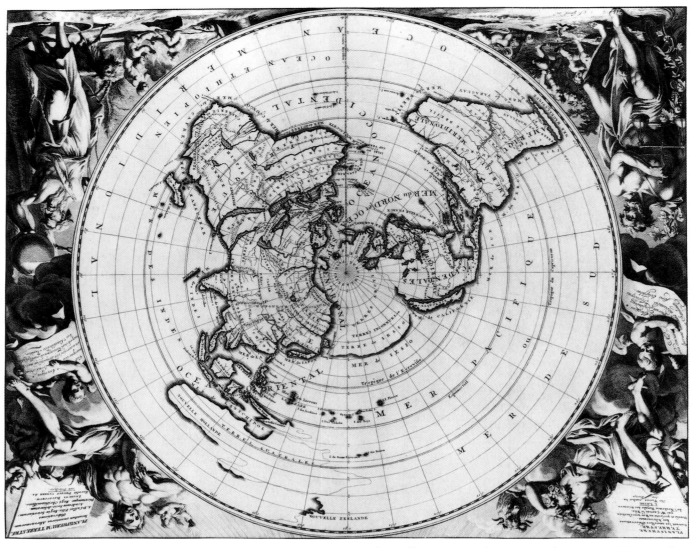

Above
Planisphere Terrestre

Pieter Van der Aa's version of an
important World map.
Leiden c1713
65 x 54cm 25.5 x 21.5in

*This large and imposing engraving is based on the map prepared
by Jean Dominique Cassini for the French Academy of Sciences
(see page 103). Cassini's original map of 1682 measured some
eight metres across, and was positioned on the floor of the
Academy; it was based on the most up-to-date astronomical
readings to determine longitudinal positioning.*

*Van der Aa's version, after a French printing of the map,
incorporates particularly powerfully engraved classical figures.*

of up-to-date reports. His maps are of great interest,
particularly in the area of North America, where he
worked in co-operation with the French. In Paris, in
1690, Coronelli issued a World map in conjunction
with Tillemont and, a year later, in Venice he
published a pair of hemisphere maps in atlas form. Both
these maps were based on details which Coronelli had
produced on globes in 1688. These large globes, of
45-inch diameter, are very rare now, as are the
complete sets of gores, but separate gores do come on
the market occasionally since they were issued in the
Libro dei Globi and, in some instances, bound into his
atlases. As Coronelli worked with the French map
makers so too did some Dutch publishers. Dutch map
publishers, many times reissuing old copperplates, were
losing their grasp on the map market, while competition from the French (1700) and Germans (from
about 1720) created a considerable balance in European
map output.

The firm of Covens and Mortier, in Amsterdam,
published some of the finest French work at the start
of the century and maintained a strong reputation
throughout. However, by about 1750 the German
producers, Homann, Lotter, Seutter and others, were
Europe's most prolific atlas publishers. World maps of
this period are not hard to find but they generally lack
the finesse of their earlier counterparts. Exceptions are

some of the maps by Sanson and Jaillot, republished in Amsterdam, and the maps of Guillaume de L'Isle, which though lacking in decoration, are finely engraved and notable for their cartographical precision.

One of the largest World maps of the early eighteenth century to appear in atlas form was that produced in England by Hermann Moll. Folded into a tall atlas folio the map measures 122×71 cm (48×28 in) and is rarely found in good condition. Published in 1724 the map is somewhat outdated (despite its author's claims), showing California as an island long after the French maps of around 1700 had depicted it as a peninsula. Jonathan Swift, it is said, referred to Moll's map in the writing of *Gulliver's Travels*, thereby adding greater credibility to his story by ensuring the location of the islands visited by Gulliver could not be disputed.

During the first half of the eighteenth century, as before, most important exploration was within the great landmasses. Surveying and map making techniques improved and the English map and chart trade, in particular, consolidated its importance. During the latter half of the century English map making was to become supreme in Europe, and the demand for spectacular World maps diminished in favour of large-scale, detailed maps of many parts of the Americas, Africa, Asia and the recently-discovered regions of Australia and the Pacific.

Below
Il Mappamondo

An elegant late eighteenth-century double hemisphere by Antonio Zatta.
Venice 1774
39 x 28cm 15.5 x 11in

This simple engraving in typical colour was issued in 'Atlante Novissimo', in Venice, by Antonio Zatta. Nearly two hundred years after its first publication the influence of Plancius's map is apparent, athough the female figures and the animals have been refined to appear less fearsome than in the original map.

Cartographically, the map shows the tracks of Cook's first voyage but presents a very distorted outline for New Zealand.

SEA CHARTS

Sea charts, whether printed or in manuscript form, often have a romantic appeal far stronger than that of land maps. The romance of the sea – man's intrepid voyages through uncharted waters and the dangers inherent in sailing unsophisticated craft even in known parts can often be sensed when examining a chart. In this chapter we look at the development of sea charts, and at some of the sea-atlases produced.

Before the development of printing there was an active chart making industry based around the Mediterranean in places like Genoa, Venice and Majorca. From such centres manuscript 'rutters' and 'portolans' were produced for the Mediterranean and Southern European Atlantic coasts. Both these terms originally referred to a sequence of written directions and sailing instructions and the expression 'portolan chart' came from the cartographic representation of such detail. The earliest known chart dates from the end of the thirteenth century and there are over one hundred known surviving charts of the fourteenth and fifteenth centuries. This number is increasing slowly as occasionally a long lost chart is found, but due to the nature of their use only a very small proportion of the original production has survived. Besides the expected 'wear and tear' on board ship, the fact that the majority of these charts were drawn on vellum (an animal skin which is particularly long-lasting) has, in certain ways, proved something of a disadvantage as once a chart was out of date the hard-wearing vellum could be put to other uses. Recently, there have been several cases on the market of fragments of portolan charts on vellum, which had been cut up and used as book covers in the sixteenth and early seventeenth centuries.

In 1485, there was the first attempt to produce working printed charts. This took the form of an *Isolario* or 'Island book', and was a combination of written directions and charts, albeit crudely engraved, of Greek islands, their ports and harbours. Bartolommeo dalli Sonetti was the author of this book and his principle was greatly expanded in 1528 by Benedetto Bordone who showed, in very crude outline, islands from all over the world including West Indian and Asian Islands and Japan. Despite the roughness of Bordone's maps the book was obviously popular, being reprinted in 1534, 1547 and around 1565.

Throughout the sixteenth century the manuscript portolan, in single sheet or atlas form, was the only relatively accurate source of navigating information.

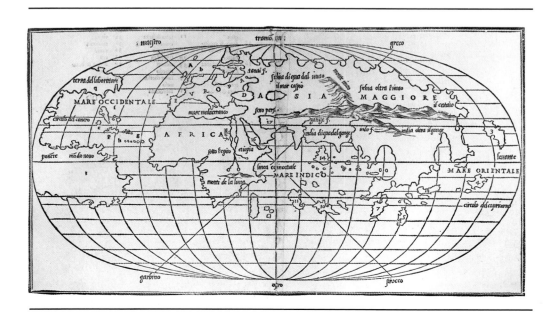

In addition to the Mediterranean schools of chart makers there were new schools in Lisbon, Dieppe, La Rochelle and London and it was these Atlantic-based centres that remained important during the seventeenth century.

By this time the Low Countries were becoming established as the European centre of map making and it is no surprise that with the expansion of their marine and mercantile empire, the production of printed sea charts should also develop here. The Mediterranean was well known and European interest was targeted beyond the Atlantic and Indian Oceans.

By 1584 an experienced seaman and pilot, Lucas Jansszoon Waghenaer, had published his compilation of charts, *Spieghel der Zeervaerdt*. In mid century the Spanish and Portuguese held strong control over maritime trade and Dutch ships became the main distributors of goods from the Iberian peninsula to other Western and Northern European ports. They had also chosen, in view of the Spanish domination in the South, to sail north into the Baltic and along the Norwegian coast to Lapland. Waghenaer had probably been on these trips and fully understood the requirements of navigators. He lived at the thriving port of Enkhuizen and, in 1579, obtained a permanent position as an excise officer. By this time his intentions to produce a chart book were materializing, and when he was dismissed from his post for certain financial irregularities in 1582, he had to put all his efforts into finding financial backing and for his project.

As soon as the first part of the book was published, with 23 charts, it was acclaimed as a great success and, together with Part II (a further 21 charts), it was republished frequently over the next 20 years. The book's success was not limited to its original Dutch text – editions in Latin (1585), English (1588), German (1589) and French (1590) soon appeared.

In view of the threat of competition from Amsterdam

Below

Zee Carte van Engelants Eyndt

Lucas Jansszoon Waghenaer's beautiful
chart of the South-West tip of England.
Leyden 1586
52 x 32cm 20.5 x 12.5in

This beautiful chart, the first of the area from the first printed atlas of European charts, extends from Plymouth to north of Mousehole and Land's End and includes the Scilly Isles (de Sorlinges). Coast profiles within the landmass add further information to the coastal detail regarding depths, sandbanks, rocks and so on.

A compass spans the mileage scale while ships and monsters decorate the seas. Busts, masks and drapes are used to elaborate the strapwork cartouche.

The large scale and robust flamboyant style of these charts, engraved by the van Doetechums, makes them visually unique in addition to their importance in the first sea atlas.

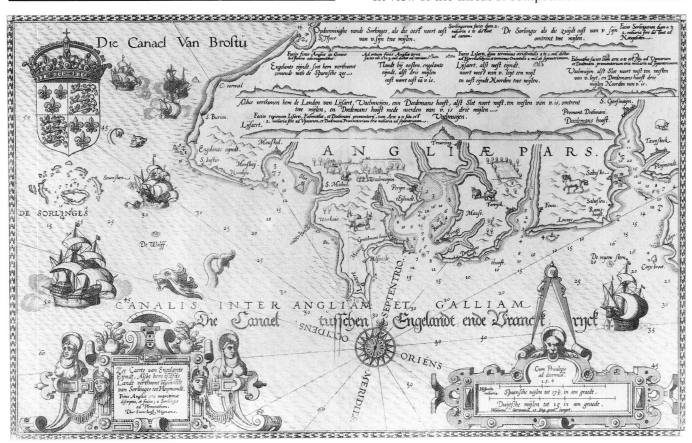

chart makers – especially Cornelis Claesz –Waghenaer soon began working on an improved chart book. The *Thresoor der Zeevaert*, first published in 1592, has more detailed charts and a different format (oblong quarto as opposed to the standard atlas folio of its predecessor). Of most interest, however, is the incorporation of coastal charts beyond the European coastline (Cadiz to N. Norway), added after the first edition. Charts of the Moroccan coast (see page 113), Trinidad and South East Asia can be found. Any edition of this work is much more rare than the earlier publication and, though Waghenaer, intended the *Thresoor* to be less 'up-market' than the Spieghel, the quality of this work and its engraving is superb.

During the first half of the seventeenth century Amsterdam saw the production of many printed charts and chart books, most of which are now very rare. The

Below

Carta Particolare dello stretto di Inghilterra tra Dover e Cales ...

Sir Robert Dudley's elegant chart of the English Channel.
Florence 1646-7
35.5 x 46cm 14 x 18in

This chart is a fine example of the series engraved by the Italian Antonio Francesco Lucini for Sir Robert Dudley's 'Dell Arcano del Mare'.

This example is a single sheet engraving whereas others, within the atlas, are double sheet. Each is simply and elegantly engraved, title

cartouches, ships and monsters, compass roses are each finely detailed. Where space allows, the engraved script is flamboyantly flourished. Despite the artist/engraver's designs the charts are well detailed and annotated – giving soundings, indicating sandbanks and safe passages and information concerning currents, winds and any other information thought of relevance to the navigator.

This is the first English sea-atlas, the first sea-atlas of the whole World, and the first on Mercator's projection. The atlas comprises 146 charts, which, according to the memoirs of Lucini, took twelve years to engrave and utilized 5000 lbs of copperplate.

Dudley had been brought up within a shipping environment – being the illegitimate son of the Earl of Leicester, brother-in-law, for a brief period, of the circumnavigator Thomas Cavendish and, consequently, well acquainted with many of the leading Elizabethan mariners. Born in 1574, by the age of twenty-one he had sailed to the West Indies where he explored part of the Guiana coast and had plundered Spanish shipping. In 1605, after losing favour at court he settled in Florence and became a Roman Catholic, from where, after about 1620 he began to put together his great sea atlas.

Right

Pascaart Veertoonende de Zeecusten van Chili, Peru, Hispania Nova, Nova Granda, en California

Hendrick Doncker's chart of Pacific coastlines.
Amsterdam 1659
54 x 43cm 21.5 x 17in

This chart, in contemporary outline colour, details California as an island (see page 160) and the supposed Pacific coastline north of Japan, the Ladrone Island group and the western coast of New Zealand charted by Tasman in 1642.

Blaeus, Gerritz, Barents and Colom, all produced notable work, but the most important publication of the period was that of an Englishman working in Italy. Gerard Mercator's new projection of 1569 had received little support from the sailors it was designed to assist and, despite Edward Wright's explanatory treatise of 1599, few maps and even fewer charts had been drawn on this principle. Sir Robert Dudley's *Arcano del Mare* was the first atlas published on Mercator's projection, the first sea atlas to cover the whole world and also the first sea atlas by an Englishman. Moreover, it is one of the finest engraved atlases of any period (see previous page). The atlas was first published in 1646–7 and a second, enlarged edition appeared, posthumously, in 1661.

Amsterdam, however, maintained its prominent position and separate charts and sea atlases appeared in profusion. In 1650 Jansson published a marine atlas as part of his *Atlantis Majoris ...* with 23 maps, entitled *Orbem Maritimum*. Work by Lootsman, Doncker, van Loon, Robijn, Roggeveen, Goos, de Wit and the Van Keulens may be encountered. Despite lack of originality in much of this work some of these productions are

technically or artistically superb (see this section and page 152 for some examples).

Designed to be sold as a supplement to Blaeu's *Atlas Major*, the *Zee Atlas* of Pieter Goos (1666–1683) is, when rarely found, often bound in vellum to match the Blaeu with charts in fine gold-embellished colour. Separate charts from the book can be found on the market and are well worth considering for their quality of design, engraving and colouring alone.

Charts were often printed on thick paper, or thin paper laid onto thick (as is frequently the case with Goos' maps) in order to prolong their working life. Another important firm of chart producers was the Van Keulen family who published charts – both manuscript and printed – and atlases from about 1680 for over one hundred years. Their dominance in the Dutch trade was confirmed by the appointment, in 1714, of Gerard van Keulen as hydrographer to the Dutch East India Company (see page 62).

During this same period both the French and the English chart making trades were developing and producing notable charts. Under Louis XIV French interest in the sciences incorporated a reappraisal of all

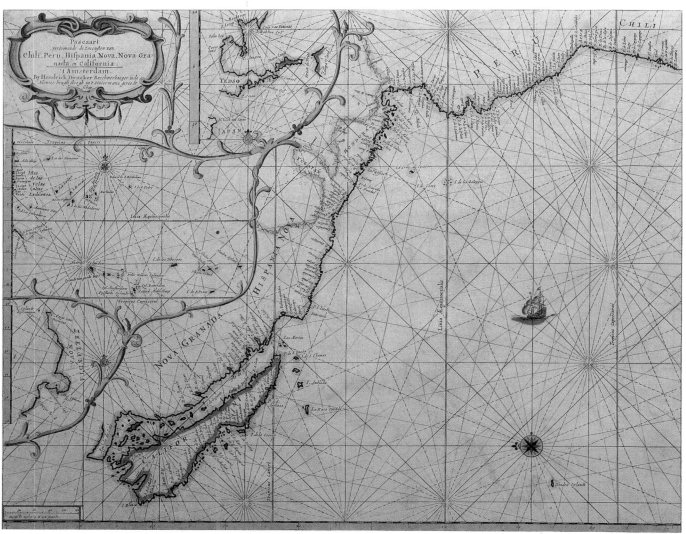

Gerard Van Keulen

The Dutch pre-eminence in map and atlas production established from the early seventeenth century, extended through certain publications well into the eighteenth. Of particular importance was Holland's production of sea atlases. Foremost of these was the Van Keulen publishing house's *Zee-Fakkel* – eventually, five published volumes charting all the seas and oceans of the world. The numerous charts in these volumes were amongst the most detailed, up-to-date and respected of their day. Editions with Dutch, French and Spanish text were published, and English and Italian versions were considered – testimony to the international demand for the publication.

The family business had been established by Johannes Van Keulen with the first atlas being published in 1680 but achieved its greatest recognition under the management of his son Gerard between 1704 and 1726. Gerard combined the essential skills of a talented cartographer/engraver with the disciplines of mathematics and an understanding of the principles of navigation. Introducing the Mercator projection into the 'Zee-Fakkel' charts, he was responsible for the production of the greatest atlas of sea charts of the time.

In addition to his family's commercial concerns, and in recognition of his skills, he was appointed Hydrographer to the Dutch East India Company from 1706. This position was one of the greatest importance since the company was responsible for the collection of detailed, secret charts and their distribution to sailors heading for the Indies. Implicitly the company not only supplied the best available information but was also kept informed of the most recent discoveries as soon as the details could reach Amsterdam.

After Gerard's death in 1726 the business was continued by various members of the family for a further hundred years, and maintained the reputation earlier established, reissuing and updating the various editions of the six parts of the *Zee-Fakkel*.

Many of the charts from these later editions show evidence of their age, being old impressions from old copper plate. Nevertheless, these rare charts are well detailed and, often, the largest scale available of any of those areas covered.

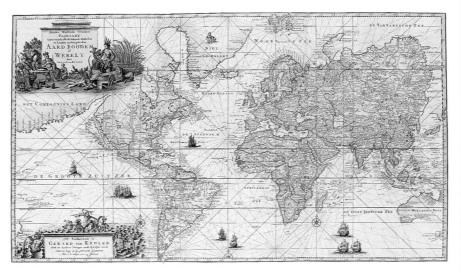

Nieuwe Wassende Graaden Paskaart ... Aard Boodem of Werelt

An attractive and important World chart on Mercator's projection by Gerard Van Keulen.
Amsterdam c1720
99 x 58cm 39 x 23in

This elegant chart is finely engraved and is in beautiful original colour with large title cartouches, ships in the seas and polar bears in Greenland. Cartographic detail is precise although the still unknown Pacific coastlines indicate various mythical land forms.

The imposing titlepiece shows natives of each continent trading. The signatory panel shows Neptune riding his chariot through the waves. This fine map was published at a time when not only was Dutch chart making supreme but Amsterdam was at its height as a mercantile and shipping power.

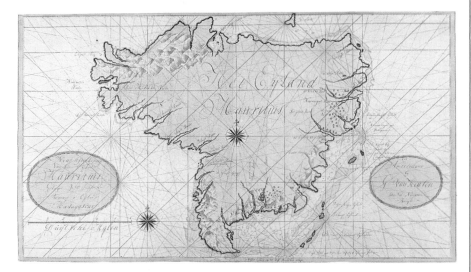

Mauritius

A fine manuscript chart by Gerard Van Keulen of the Indian Ocean island.
Amsterdam c1720
99 x 59cm 39 x 23in

Attractively designed and excellently executed, this manuscript has the signature of Gerard van Keulen and would have been amongst those made available, by special request, under strict control.

No printed version of this chart exists and, although the island detail is minimal, the attention to detail around the coasts, and particularly the harbours, is remarkable.

The island of Mauritius was of great importance as a stop-off point to or from the East Indies, and this is a fine example of the combination of artistic skill and surveyor's detail, using pen, ink and delicate wash colours.

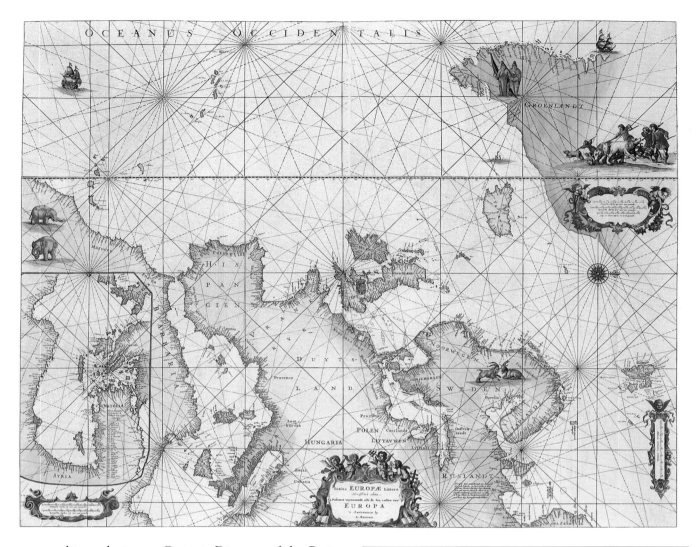

Above
Totius Europae Littora

Louis Renard's issue of a fine European
chart.
Amsterdam c1675–1715
88 x 72cm 35 x 28.5in

*Around 1675 Frederic de Wit had published his 'Atlas Maritime'.
Forty years later Renard had acquired and revised the plates and
issued them under the title 'Atlas de la Navigation'.*

*De Wit's original European chart had some additional coats
of arms but lacked the westward extension of the chart which
includes the Azores and Greenland. Renard had this sheet
engraved and added to the original.*

cartographic techniques. Cassini, Director of the Paris Observatory, had initiated and seen the completion of a new survey of France from which new coastal charts were prepared and published in 1693 in Paris and Amsterdam by Hubert Jaillot in conjunction with Pierre Mortier. The first edition of *Le Neptune François* included other European coastlines in 29 large charts. Frequently found bound in combination with *Le Neptune* are the nine superbly engraved charts (amongst the most elaborate of any period) by Romein de Hooghe (see page 64), and the *Suite du Neptune François*, a collection of 37 charts of foreign waters published by Mortier. In this, charts of the African, Asian and North American coastlines are well represented (see page 20). The original *Neptune François* plates eventually found their way to the official French *Dépôt des Cartes et Plans de la Marine* at Versailles and from here, in association with maps by Jacques Nicolas Bellin, were reissued from 1753 until into the next century.

Manuscript work by the London chart makers of the Thames School had been respected and employed since the early seventeenth century and by around 1670 the printed charts and sea atlases of John Seller and, subsequently, John Thornton and others were widely used.

In 1693 the first national survey of the British coasts, by Captain Greenville Collins, was published in London (see page 85) and its life extended for nearly one hundred years. However, of far greater importance

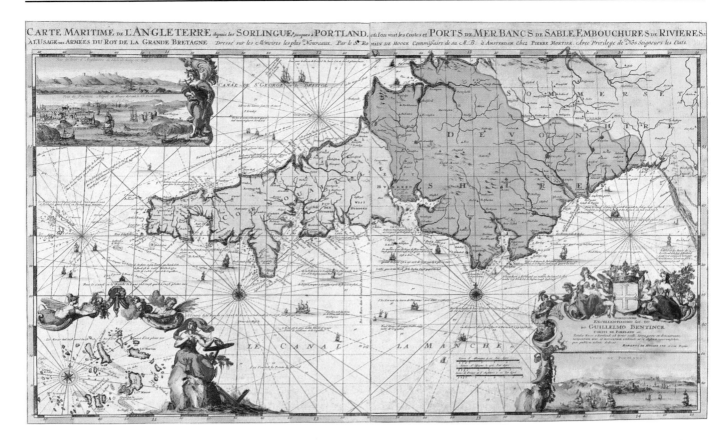

Above
Carte Maritime de L'Angleterre ...

A magnificent chart of England's south-
west coastline by Romein de Hooghe.
Amsterdam 1694
95 x 58cm 37.5 x 23in

*This exceptional chart, in original colour, is one of nine which
were designed and engraved by Romein de Hooghe for inclusion
in Pierre Mortier's 'Neptune François'. Covering the European
coastlines from southern England and Holland to include the
entire Mediterranean these are amongst the most spectacular of
any engraved sea charts.*

during this period were the activities of numerous
English chart makers active along foreign shores whose
work was subsequently published in London.

John Seller in his *Atlas Maritimus* of 1675 had relied
extensively on earlier Dutch work, either using the
original copperplates or copying directly, with the
exception of only a few charts. However, some of these
'new' charts are of particular interest, showing the latest
reports from the new English colonies on the American
north-east coast. Through the publications of John
Seller, his colleague John Thornton and others, we can
trace the origins of one of the most important
eighteenth-century chart books. The *English Pilot*,
whose first volume appeared in 1671, was ultimately
composed of six parts, each containing between twenty
and thirty-six charts, covering the navigation of each

section of the World. The *Pilot* was published in
numerous editions throughout the eighteenth century,
by the prolific chart making firm of Mount and Page.

From around 1750 increasing numbers of better
quality charts were produced of the China Seas and of
the North American coasts. English surveyors and chart
makers worthy of note include William Herbert, Joseph
Speer, Alexander Dalrymple, William Heather, Captain James Cook, Robert Sayer and John Bennett.
French chart producers include Le Rouge, d'Après de
Mannevillette, Dumont d'Urville and Sartine.

One of the greatest sea atlases of any period is Des
Barres' *The Atlantic Neptune*. Covering only the North
American Atlantic coasts this is described in the
chapter on North American maps (see page 157).

As with the increasing sophistication of land
cartography so, by 1800, European marine hydrography
was moving into the modern phase. The British
Admiralty had established, in 1795, its own Hydro-
graphic Office under the supervision of Alexander
Dalrymple, and its first charts were published in 1801.
French marine explorers in Cook's tracks were active
in the Pacific, and were supplied with charts by the
Dépôt des Cartes et Plans de la Marine. In the United
States, a *Survey of Coasts* was initiated in 1809,
although little happened until the 1830s when chart
production commenced.

Many of these nineteenth-century charts remain the
base models, with few revisions, for today's sailing
charts.

TOWN PLANS

This chapter concentrates on town plans issued in general atlases of a non-national, i.e., world-wide content. We shall look separately at British town plans beginning with those by John Speed, and deal with town plans from travel books as opposed to atlases, within their relevant geographical section.

Town plans are designed to provide the viewer with a more precise awareness of a particular place than can be shown on a map of a larger area. Not only does the large scale of a town plan allow the identification of small details but the map maker/engraver can also give specific information on local customs, local economy and the actual appearance of buildings.

The greatest publication in this genre is undoubtedly the Georg Braun and Frans Hogenberg's *Civitates Orbis Terrarum*, published in Cologne from 1572. This great work is a combination of three specific styles of town representation, each with its roots in Medieval art:

1. *The Panorama* A purely pictorial view of a town showing buildings in two dimensional profile;
2. *The Plan* A scaled representation as viewed from directly above; a large scale map or plot of the land –an extension of the surveying technique for estate plans or building or engineers' plans;
3. *The Bird's-eye* An oblique view – a combination of both panorama and plan. This innovation permitted both cartographic and pictorial detail to be shown and proved enduringly popular. Before the *Civitates* there had been only a very few printed plans produced, apart from some by Sebastian Munster of German cities in his *Cosmographia*, and by Lodovico Guicciardini in his *Descrittione di tutti i Paesi Bassi* of Low Country towns. Also mentioned are the town panoramas in Hartmann Schedel's *Nuremberg Chronicle*, in which 26 double page and numerous single page boldly woodcut views of European towns can be found (see page 108).

The *Civitates* was compiled and written by Georg Braun, Canon of Cologne Cathedral. Braun gathered together vast amounts of information and draft plans to produce over 500 city views/maps published in six parts between 1572 and 1617. Most of these engravings were made by Simon Novellanus and Frans Hogenberg, many after drawings by Joris Hoefnagel, a professional topographic artist. Hogenberg, who had also produced maps for Ortelius's *Theatrum*, died in about 1590 and

his work was continued by one Abraham Hogenberg, presumed to be his son.

The majority of views in the book are, naturally, of European cities with a particular emphasis on the Low Countries and Germany, but there are also views of Middle Eastern cities, North and West African ports, Indian Ocean ports, Cuzco and Mexico. This great work, in the systematic tradition of Ortelius, is one of the great books of the Renaissance, and among the loose sheet maps which now come onto the market, often in fine original colour, collectors have the opportunity of finding the earliest plan of their town or city.

After the book's last edition in 1618, the copper-plates remained in Cologne until about 1653 when they were acquired by Jan Jansson and reissued four years later with reset text. Some of the plates were subsequently reissued by Jansson's heirs in 1682, by Frederic de Wit in 1710, and by Covens and Mortier later in the century. The London map, for example, was republished by Jansson and again in 1708 by Christopher Hatton in his history of London; in each case the copperplate suffered some re-engraving.

In the absence of re-engraving of the copperplate, and without means of checking text on the reverse of the map to identify the edition, one can assume that the original Cologne issue of a plate was printed on smaller size paper than the later Dutch editions.

Imitations

The success of the *Civitates* led to a variety of imitations – later editions of Munster's *Cosmographia* included copies of many of the plans. François de Belleforest incorporated copies of the *Civitates* originals in his own edition of Munster's work. Francesco Valegio in Venice in 1595, and Daniel Meissner *c* 1625, issued miniature town books based almost entirely on the *Civitates*, while the town views and plans seen in the borders of most 'cartes à figures' (see page 27) can be traced back to the same source.

Amstelredamum ...

Georg Braun and Frans Hogenberg's detailed plan of the City which was to assume prominence in the history of cartography.
Cologne 1572
49 x 34cm 17.5 x 14in

A detailed 'bird's-eye' plan showing buildings in profile and numerous ships at anchor. It is noticable how the pattern of sixteenth-century Amsterdam is clearly recognizable today.

This plan was taken from one by Ludovico Guicciardini – an Italian engraver who had supplied the maps for 'Descrittione di Tutti i Paesi Bassi', first published in 1567.

The Civitates comprised some 548 views and plans, the majority being of Germany and the Low Countries. The composition, apparently dependent on availability of source material, as opposed to the editor's requirements, was as follows:

British Isles – 20
Iberian peninsula – 49
France – 39
Belgium – 48
Holland – 55
Germany – 118
Switzerland – 18
Austria – 15
Bavaria – 9
Bohemia and Moravia – 11
Hungary, Croatia and Transylvania – 19
Denmark and Norway – 30
Sweden – 5
Poland – 13
Russia – 3
Italy – 58
Mediterranean and Near East – 12
Asia – 6
Africa – 18
America – 2

These appeared in six volumes, whose first issue dates were 1572, 1575, 1581, 1588, 1598 and 1617.

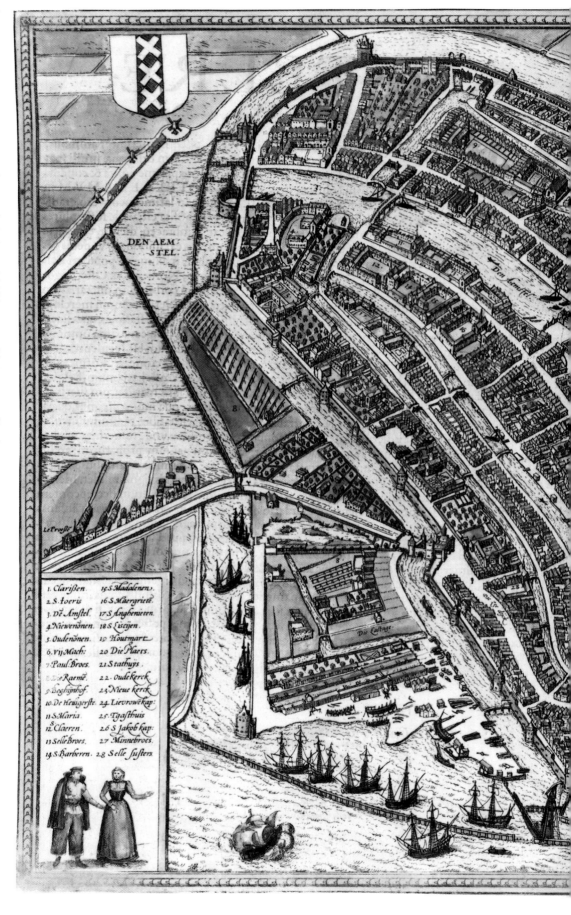

AMSTELREDAMVM, nobile Inferioris Germaniæ oppidum, ad recipiendos, ex omnib. mundi partibus, mercatores, recenter natum, genus hominum incolit mercimonijs deditum quæ quidem, tum blanda populi comitate, ac sedula diligentiâ, industria; tum portus commoditate permagna, usq, adeo incrementa sumpserunt, vt vix vllum mercaturæ genus excogitari possit, quod hîc non exerceatur. Hinc fit, vt opum lucriquè cupiditas, ex remotißimis etiam terris, negotiatores, in hanc ciuitatè inuitet, qui varia hinc bona, & maximè, rem frumentariá, in Brabantiam, cæteraq, longè dißita, totius vniuersi loca, transferentes; ingentes ex eiuscemodi commertio opes consequuntur.

Die Braeck

Du Kloek

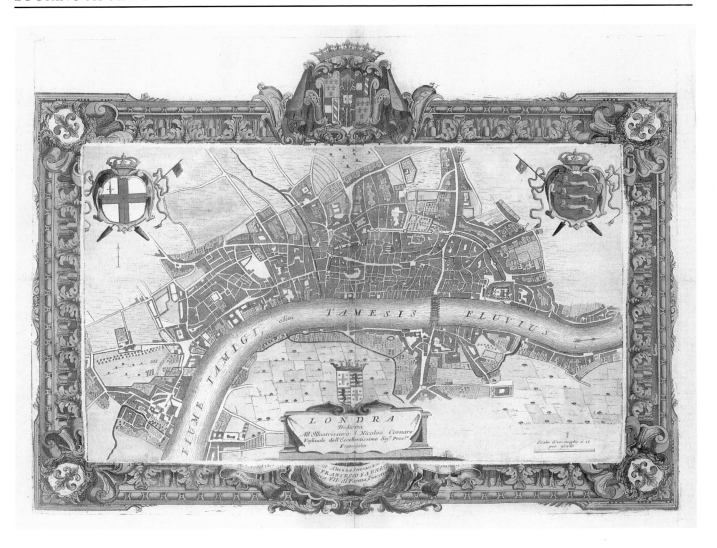

Above
Londra

A decorative map of London by Vincenzo
Coronelli.
Venice 1689
52 x 41cm 20.5 x 16in

*Besides Francesco Valegio's miniature copies of the Braun and
Hogenberg originals, there were no series of town plans issued in
Italy until those of Vincenzo Coronelli, appearing in a variety of
formats and books on specific regions.*

*The map is detailed and well engraved but lacks street names.
The English coats of arms, and that of the City of London decorate
this map, which is particularly unusual in being printed in two
separate 'pulls'. The map is a separate copperplate from the very
ornate frame and appears more often without the engraved
frame.*

In Holland Jan Jansson acquired the *Civitates*
copperplates to contest the market with Joan Blaeu,
who from 1649 had published his own collection of
around 220 town plans of the Netherlands. Additionally, from 1663 Blaeu issued three town books of Italy
– of the Vatican, of Rome, and of Naples and Sicily
(74, 41, and 33 plates respectively) – and, from 1682,
Blaeu's heirs issued books of Piedmont (69 plates) and
Savoy (71 plates). These copperplates, as with the
Jansson, Braun and Hogenberg plates, finally came into
the hands of de Wit and Pierre Mortier.

During the early eighteenth century in Amsterdam
reprints of these earlier copperplates were still being
published and, in Leiden, Pieter Van der Aa was
publishing view books, primarily though, of panoramas
rather than plans.

Other notable publications

In Germany there are three publications worth noting:
Matthaus Merians' (father and son) prolific output of
travel books illustrated with maps and panoramas from
about 1630 and, about one hundred years later, the
atlases incorporating town plans by Johann Baptist
Homann (see page 109) and George Matthaus Seutter.

Vincenzo Coronelli in Venice, *c* 1690, had published
a number of localized view books and these invariably
included town plans (see above). Over the next one
hundred years each European country produced its own
localized mapping. However, the collector is most
likely to find plans from the following publications:

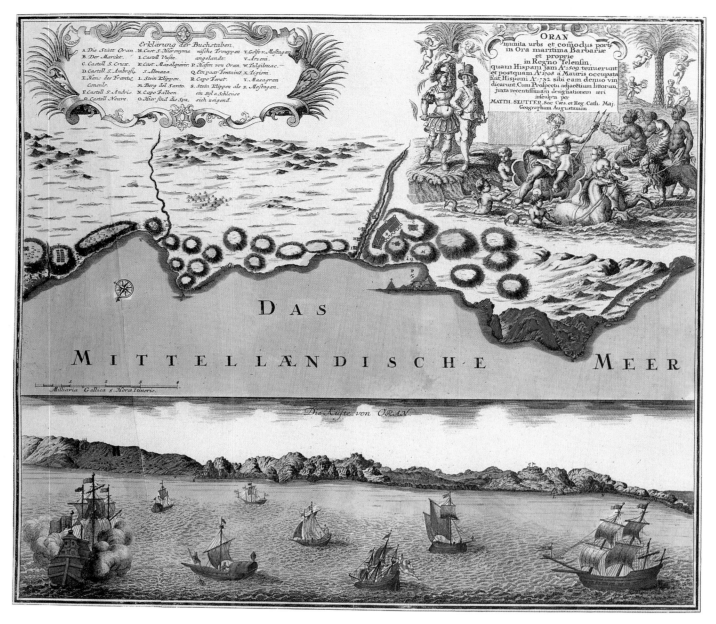

1764 The *Petit Atlas Maritime* by Jacques Nicolas Bellin included amongst 582 maps and charts, plans of most of the World's major river and coastal ports. Paris.

1771 John Andrews' *Collection of Plans of the Capital Cities of Europe and some remarkable Cities in Asia, Africa and America*. Some 40 plans including New York, Iedo (Tokyo) and other Asian and European cities. Published in London from 1829. A collection of maps and some plans published under the 'Superintendence of the Society for the Diffusion of Useful Knowledge', (i.e., S.D.U.K.) – detailed maps of many European, some Asian and some North American towns.

Besides those town plans which were specifically designed as such, many plans, as we shall see, including John Speed's of the English, Welsh and Irish Counties, were incorporated as insets on more general maps. Despite their apparent secondary importance these are often the first plans of those cities shown and should not be ignored by collectors of town plans.

Above
Oran

A decorative and unusual plan and panorama of this Algerian city by Georg Matthaus Seutter.
Augsburg *c*1750
54 x 49cm 21 x 19in

This bold engraving in original colour typifies the style and format of the series of bird's-eye plan/maps, in association with panorama/profiles, of major cities of the World published in Germany during the eighteenth century.

In addition to those plans issued in atlases and books, there have been from around 1750 large numbers of separately issued town plans published for practical use. Loosely folded into covers or slipcases, mounted on heavy paper or cloth, or even printed on silk, these everyday accessories were especially vulnerable, hence their proportionately small survival rate.

The Ptolemaic British Isles

The Ulm edition of Ptolemy in superb
original colour.
Ulm 1482
46 x 39cm 22.5 x 15.5in

This spectacular medieval map illustrates the Ptolemaic outline
of the earliest printed maps of the British Isles. The Ulm edition
of Ptolemy was the first published north of the Alps and the first
using woodblock printing – it is also the first of any map series
which is likely to be found in contemporary colour. The strong

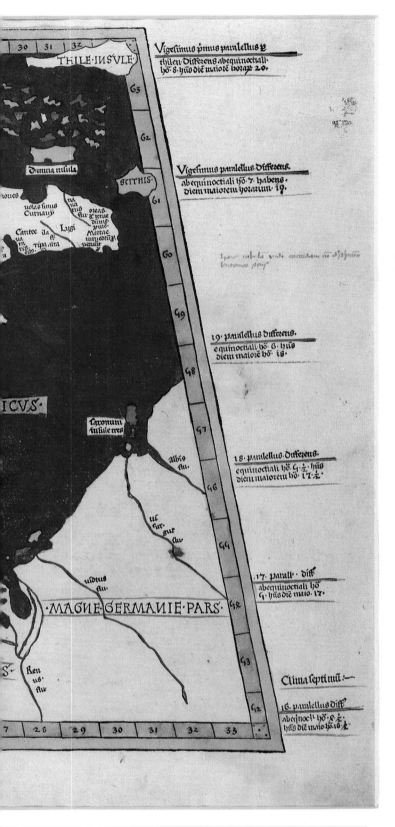

colours here typify the first edition compared with the softer colours of the second edition of 1486 (see page 118).

BRITISH ISLES

Maps of our home countries are encountered in a variety of compositions: the British Isles encompassing the whole group of islands; Britain, usually shown as England, Scotland and Wales; the countries of Wales, Scotland and Ireland shown separately; sectional regional maps often taking no account of national boundaries; and finally maps of individual counties within each national region.

For convenience therefore, within this large category, I have divided into a number of subheadings: General maps; England and Wales; British Town Plans; Scotland and Ireland.

General maps

Thirty-four years after the first issue of Ptolemy's map of the British Isles with its pronounced East-West alignment of Scotland, Bernard Sylvanus, in Venice in 1511, had published his edition of *Geographia* which was innovative in a number of ways. Firstly, the maps, printed in black, had letterpress printed by moveable type with the more important names printed in red; the atlas was therefore the first to be printed in two colours. Secondly, Sylvanus had redrawn the Ptolemaic outlines and added a completely new World map to the traditional geography (see opposite). Sylvanus' map of the British Isles was the first to correct the typical West-East Scottish slant, seen on all previous Ptolemaic maps, and was followed two years later, in 1513, by Martin Waldseemüller's *Tabula Nova ...* (see page 74). This map was a further improvement and added more names, particularly in the south coastal areas of England and Ireland, regions visited by merchants and fishermen from the European mainland and, consequently, best known to the portolan chart makers from whose work these maps were compiled.

Through the sixteenth century, maps of the British Isles became more refined and improved rapidly. Sebastian Munster's woodcut map of 1540 was a marked improvement, followed in 1546 by George Lily's magnificent copperplate map of the British Isles – the first separately-issued map of the islands and one of the most elegantly engraved maps of the period. There were a number of later versions of the Lily map and all are rare.

Until 1570, when Ortelius and Gerard de Jode issued copies of a new map by Gerard Mercator, printed maps of the British Isles were either of the Ptolemaic,

Sylvanus/Waldseemüller or Munster/Lily format. In 1564, Gerard Mercator had engraved a fine new map which, unfortunately, is now little known and believed to exist in only three copies. However, the map was readily available to contemporary map makers including Ortelius and de Jode who produced versions for their respective atlases. The Ortelius version is undoubtedly a finer engraving and visually more pleasing, while the de Jode is appreciably more rare. The former was issued in over 30 editions over the 42-year life of the *Theatrum*, the latter in only two editions of 1578 and 1593, in addition to its separate publication in 1570.

Included in Ortelius's atlas from 1573 to 1602, besides a historical map of the British Isles from 1590 on, were Humphrey Lhuyd's maps of England and Wales, and Wales alone. Lhuyd, a Welsh historian and topographer, had died five years before publication of his work, which in 1603 was replaced by a version of Christopher Saxton's fine map of England and Wales. First issued in his atlas of 1579, Saxton's map became one of the most influential of this sequence of British maps and was copied by numerous other map makers both in England and on the Continent. The Ortelius issue, however, only appeared in one edition of the *Theatrum ...* and is, consequently, rare being replaced in 1606 with a spectacular map by Jan Baptist Vrients. This map includes Ireland, is finely engraved and is notable for the genealogical table of the Royal Family.

From the start of the seventeenth century some fine maps were produced in London and Amsterdam. These maps appeared not only in atlases of the British or English counties but also in world atlases and there is a plentiful variety for the collector or buyer of decorative pieces. The shape of the British Isles lends

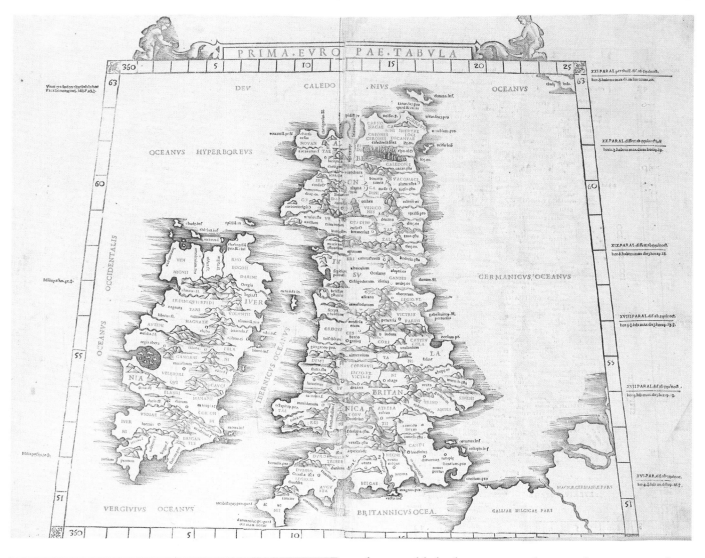

Above
Prima Europa Tabula

Bernard Sylvanus' improved outline of
the British Isles.
Venice 1511
41 x 34cm 16 x 13.5in

*This interesting and rare map by Bernard Sylvanus represents the
earliest attempt, in a printed map, to remove from the Ptolemaic
standard as shown in previous issues of the 'Geographia'. Here
Ireland is shown in far truer proportion and position than before,
the earlier eastwards slant of Scotland is corrected and more
placenames are given (see pages 70 and 71).*

*This particular publication is important as the first use of true
colour printing on a map – a practice which appears most
effectively here, the more important names being printed with red
inked letterpress, but which was repeated only rarely by other
map makers. This issue of 'Geographia' in fact was only published
in one edition.*

those published maps in atlases, others appeared in
history and geography books, or were separately issued
as wall maps for contemporary decoration, functional
or educational use. To list particularly desirable maps
would mean omitting some without due cause other
than lack of space, so I suggest the reader studies the
illustrations to appreciate the range available.

In addition to the strictly topographical content of
most maps, collectors will be able to find maps which
illustrate particular 'themes', including such categories
as: historical maps; maps with roads, canals or railways;
geographical or physical maps; and so on. These
'thematic' maps, produced mainly in the nineteenth
century, are dealt with in a later chapter (see page 177).

itself to fine map design, allowing the artist/map maker
to include elaborate title and dedication cartouches,
inset views, plans or panoramas, or vignettes of ships,
monsters, figures and historical scenes. In addition to

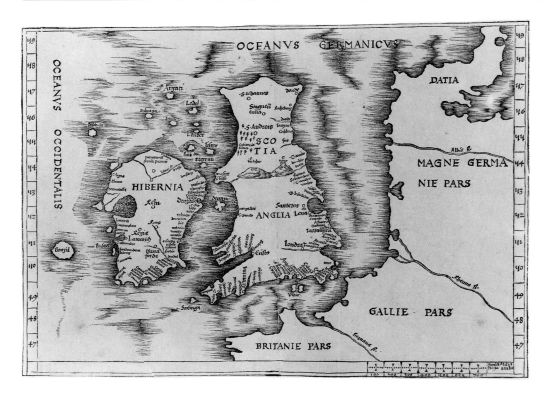

Left
**Tabula Nova Hibernie
Anglie et Scotie**

Martin Waldseemüller's
influential map of the
British Isles.
Strassburg 1513
51 x 38cm 20 x 15in

*After ten editions of Ptolemy's
'Geographia', Waldseemüller's
edition marked the transition
from the ancient to the new
world in maps. For this map
Waldseemüller would have
drawn on contemporary manuscript charts and reports to
produce the best map of the
British Isles to this date. This
map, the standard for the
next thirty years, was replaced by the maps of Munster and Lily.*

Angliae Regni ...

Ortelius's map of England and Wales
supplied by the first Welsh cartographer,
Humphrey Lhuyd.
Antwerp 1573
47 x 38cm 18.5 x 15in

*On this fine map, in original colour, Lhuyd receives full credit for
his authorship. Ortelius also used his map of Wales in the atlas.*

Above
Anglia et Hiberniae

Johann Baptist Vrients' replacement for
Lhuyd's old map.
Antwerp 1605
57.5 x 43.5cm 22.5 x 17in

*This beautifully engraved map, here in fine original colour, is one
of the most sought after of all maps of Britain.*

*Incorporated into the Ortelius atlas by its publisher Vrients after
1606 and used subsequently, the map is detailed, covering
England, Wales and Ireland and has a distinctive genealogical
table of English Kings and Queens since 1066. Ships, a monster,
a mermaid, and Neptune decorate the blank areas of sea.*

*This map, being from a larger than average copperplate, is
frequently damaged owing to the necessity of folding into the atlas,
or because the paper margins are so close to the printed surface.*

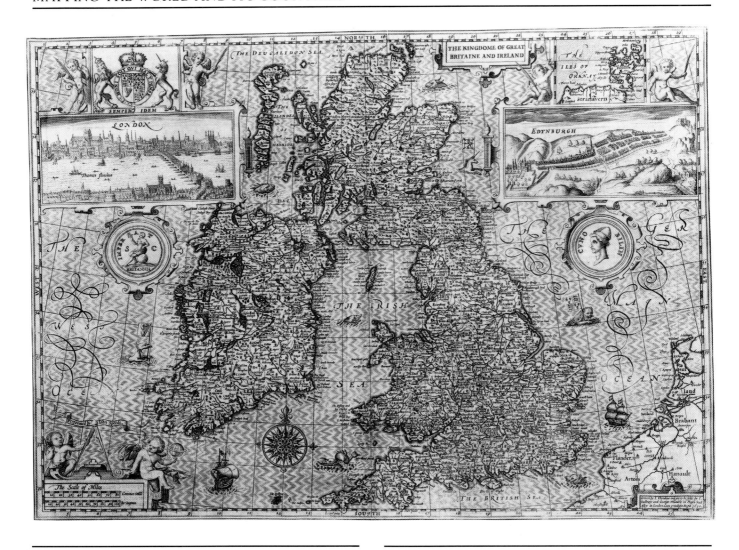

Above

The Kingdom of Great Britaine and Ireland

John Speed's famous map of the British
Isles.
London 1611
51 x 38cm 20 x 15in

This finely engraved map is the work of Jodocus Hondius who engraved all the other maps for Speed's 'Theatre of the Empire of Great Britaine and Ireland …'. The general maps, however, are generally better designed and more carefully engraved than the county maps. Here John Speed, whose name appears in a banner at bottom left, has incorporated the cartography of Saxton for England and Wales, Mercator for Scotland and Hondius for Ireland.

The views of London and Edinburgh are taken, in the first instance, from a drawing of c1600, later engraved and published by Visscher, and in the second, from a drawing of Edinburgh in 1544.

Opposite, top

Britannia

Joan Blaeu's version of an earlier map by
John Speed.
Amsterdam 1645
53 x 42.5cm 21 x 17in

This famous and exceptionally decorative map is instantly recognizable for the two broad columns showing, on either side, seven vignette portraits of early Saxon kings and seven vignette scenes from the lives of the later Saxon Kings.

The title of John Speed's prototype map –'Britain As It Was Devided in the tyme of the Englishe-Saxons especially during their Heptarchy' – describes fully the cartographic content of the map which was copied by Blaeu and, a year later, by Jansson. The Blaeu issue, illustrated here in fine original colour, is probably the most rare of the three versions. Jansson's map, issued until the mid-eighteenth century, is often mistaken for the Blaeu but can be distinguished by the presence of compass or rhumb lines.

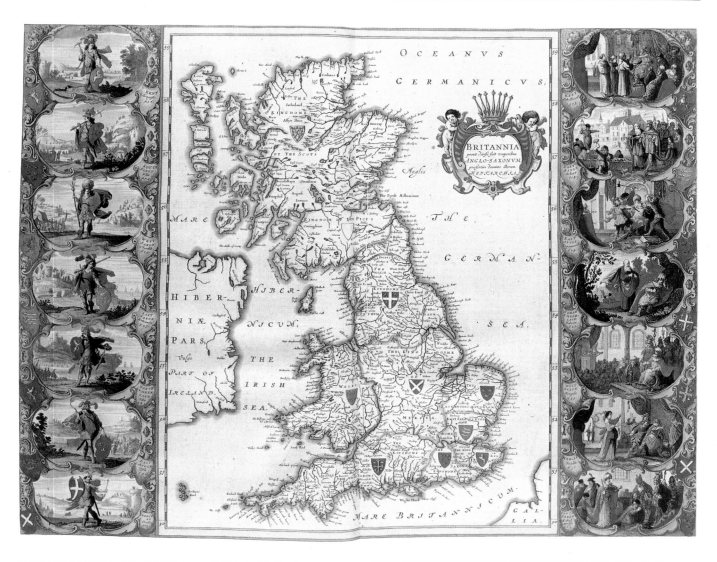

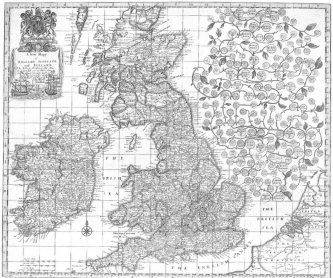

Right
A New Map of England, Scotland and Ireland

Robert Morden's interesting and rare
map.
London c1680
58 x 51cm 23 x 20in

The Royal coat of arms surmounts the titlepiece of this map, whilst below a scene shows a naval battle raging – an indication of Britain's expanding maritime interests. A delicately designed genealogical tree displays British Kings and Queens.

Robert Morden is best known for his county maps which were published in the 'Britannia'. However, he also produced a large number of single sheet folio size maps of all parts of the World but these were not issued in atlas form and so are consequently scarce.

ENGLAND AND WALES

From 1579 when Christopher Saxton published his atlas of England and Wales – the first national atlas – the majority of county atlases produced in England and in Amsterdam included maps of each individual county. For the serious collector of county maps up to 1850 there are upwards of one hundred notably different maps of each county and, if one wishes to go into further detail, many variations of states and editions of these maps. Given such a variety, I list below, with notes, some of the map makers whose work is most in demand and whose work is most likely to be encountered.

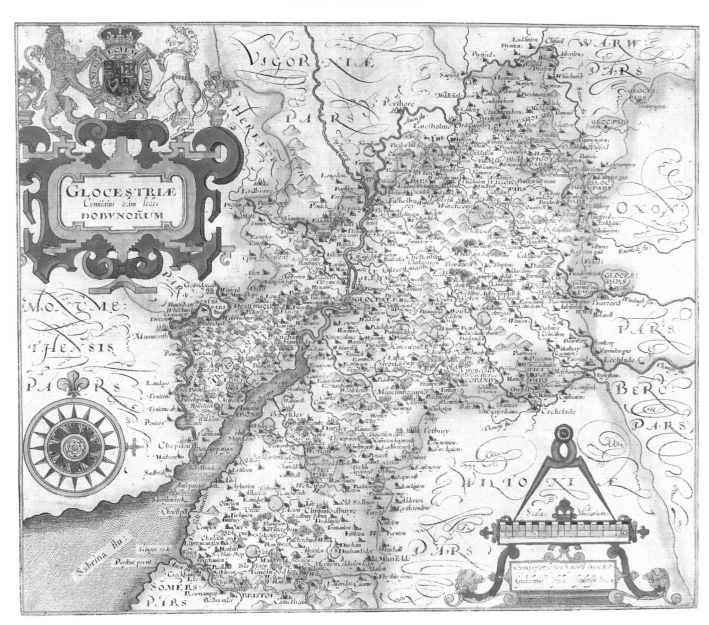

Christopher Saxton

Born in Yorkshire in 1542 or 1543, Christopher Saxton is, with John Speed, one of the best known figures in English cartography. In his early life he was fortunate enough to have gained an appreciation of maps and map making through the enthusiastic advice of the Saxton family's local vicar, John Rudd of Dewsbury, himself a cartographer. In England, as on the continent, the period was one of expansion and enlightenment and at the forefront of the newly practised arts and sciences stood surveying and map making. Saxton committed himself to surveying and saw printed and published, in 1579, the county atlas whose maps were to be the foundation for the mapping of English and Welsh counties for almost two hundred years.

Little is known of Saxton's early life until around 1570, when under the patronage of Thomas Seckford, lawyer and official in the Court of Queen Elizabeth I, he is found working on the series of maps which had been commissioned by Seckford. By 1574 the first map plates were engraved and five years later the atlas was ready for publication. The fact that the atlas of all England and Wales was prepared so quickly is remarkable given that Saxton himself is credited with the surveying of each county. However, it must be assumed that he drew, wherever possible, on extant work. Conceivably, Mercator's map of the British Isles (see page 72) might have been used and there is some slight evidence to suggest that this map was itself based on information supplied by the vicar, John Rudd.

Saxton's atlas comprises a general map of England and Wales and 34 others of individual or grouped counties. These were engraved by a number of the best engravers of the period and, when found in the original colour, compare favourably with the finest maps produced at any period in Europe.

Saxton's atlas was reissued, with maps revised and altered, several times over the next two hundred years and, apart from a general map of England and Wales, and one of Wales, is Saxton's only printed contribution (now known) to cartography. However, he continued his career as a surveyor and a number of estate plans, particularly within the northern counties, are known to be the work of both him and, subsequently, his son Robert.

Left
Glocestriae

Williams Hole's re-engraved and reduced version of Christopher Saxton's map of Gloucestershire.
London 1607
31 x 28cm 12 x 11in

This charmingly engraved map is based on the first engraved map of the county published originally in 1579. Williams Hole and Kip re-engraved Saxton's original maps for the first issue of Camden's 'Britannia' to include county maps. In many instances – notably Surrey, Middlesex, Kent, Sussex; Oxford, Berkshire and Buckinghamshire; Northampton, Bedfordshire, Cambridgeshire, Huntingdon and Rutland; Warwick and Leicestershire; Lincoln and Nottingham; Westmoreland and Cumberland; Denbigh and Flint; Montgomery and Merioneth; and Radnor, Brecknock, Cardigan and Carmarthen; counties were grouped together as one by Saxton. Consequently, many of the Britannia maps are the earliest individual county maps available now to collectors.

Most of these maps give credit to Saxton in their titles and, as such, are often referred to as 'Saxtons'. Such a misplaced naming frequently leads to disappointment since the price differential between the originals and the much more common Kip and Hole issues is, on average, ten-fold.

The three editions of these maps are identifiable as follows … 1607 – maps have Latin text on the reverse; 1610 – have no text on the reverse; 1637 – have the addition of an engraved plate number in the lower left corner of each map.

Above
Detail from 'Middlesex' map

From Camden's 'Britannia'.
London 1607
Overall map size 33 x 27cm 13 x 10.5in

This section of the Middlesex map from the 'Britannia' shows the Cities of London and Westminster and the country round about.

Although the engraving of detail is somewhat representative, such landmarks as London Bridge, St Paul's Cathedral, Southwark Cathedral and the Palace of Westminster can be identified. Note the gallows at the corner of Hyde Park. Maps of this period did not show roads. The ornamentation at the lower edge of the illustration is that of the top band of the key.

Although there is no engraver's signature on this plate one can assume, through similarities of style, that either Kip or Hole was responsible for the work. However, it is interesting to note that the original of this map was not by Saxton – John Norden, a surveyor from Middlesex, produced maps of Hampshire, Surrey, Sussex, Kent, Hertfordshire and Middlesex, which were regarded as of better quality than Saxton's and these were preferred for use in this book.

After Saxton

Often described as 'Saxton' maps, since his name appears on most, the charming maps engraved by Williams Hole and Kip appeared in the 1607 edition of William Camden's *Britannia*. This popular history and guide was first published in 1586 without maps and subsequently with maps in 1607, 1610 and 1637. The maps, all appreciably smaller than the original folio size, are based primarily on Saxton, or where available, on the originals of John Norden. Given the rarity of the original Saxton editions and a series known as the 'Anonymous' maps published around 1602 (also rare and covering only 12 counties), these are the earliest maps of individual English and Welsh counties now available to collectors.

John Speed

Arguably the most famous name in map making, certainly for British maps if not the world, is John Speed. His maps of the counties of England and Wales and Irish regions are unique in their combinations of topography, history and cartography, incorporating not only a wealth of information relating to that county but often town plans made by Speed himself. Speed,

Above
Hartford Shire

John Speed's fine map of the county of
Hertfordshire.
London 1610-11
50 x 37cm 19.5 x 14.25in

This map, in fine original colour, is amongst those of the English and Welsh counties published in the 'Theatre of the Empire of Great Britain'. This map is a fine example of those produced by Speed and engraved by Jodocus Hondius (whose signature appears next to the publisher's imprint at lower left).

Surrounding the detailed map is a wealth of information; in the top left corner is a plan of Hertford and, at top right, a plan of Verulamium and a view of St Albans; to the right of the map are battle scenes and large, ornately framed panels of text describing the county's history; to the left are the coats of arms of 'Roger Erle of Clare and Harford' and 'Edward, Seinor Erle of Hartford' and a muse studying a globe.

The most commonly seen edition of John Speed's maps are those prior to 1676 – the so-called 'Basset and Chiswell' edition. However, although the very late editions of John Speed's original copperplates are more scarce, traditional demand has been for the earlier, stronger-impression editions where the engraved line can be clearly seen, as opposed to the later editions.

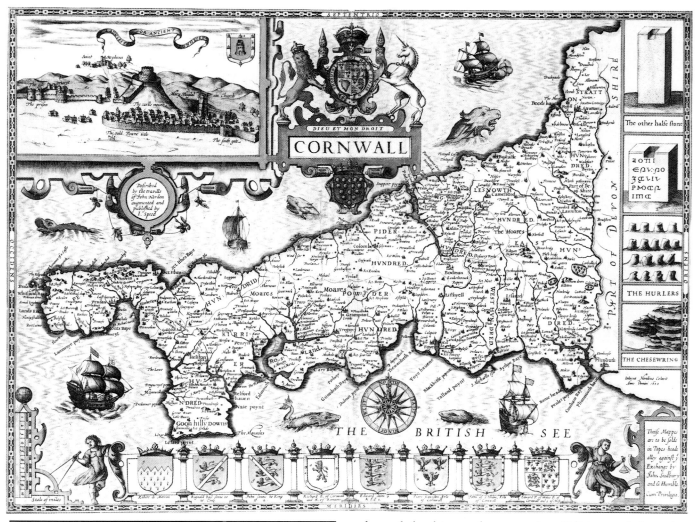

Above
Cornwall

By John Speed. One of the most famous
and identifiable maps of the westernmost
English county.
London 1610-11
51 x 38cm 20 x 15in

*A superb map which the engraver, Jodocus Hondius, has
crammed full of information and decoration. The view, top left,
is of the ancient town of Launceston, and, at the right, are panels
of natural features and archaeological remains. Coats of arms of
the county's noblemen decorate the lower border while the Royal
arms are shown above the title.*

*This is one of the few Speed maps where an alteration was
made to the copperplate. Accordingly, editions after 1623 have
the words 'The Irishe or Virginian Sea' set in the sea area between
the lower frame of the panoramic view and the coastline north
from Land's End.*

that of the history by centuries. Editions of the atlas
appeared in 1611, 1614 and 1616 (with Latin text, as
opposed to English on the verso, to appeal to the
Continental market), published by John Sudbury and
George Humble; in 1623, 1627 and 1631–2 by George
Humble alone; in 1646 and 1650–54 by William
Humble; 1662–65 by Roger Rea; in 1676 by Thomas
Basset and Richard Chiswell; c 1690–95 by Christopher
Browne; c 1710–45 by John and Henry Overton; and
c 1770 by Cluer Dicey and Co. Edition identification
of individual maps may be made by checking the
publisher's imprint, the text setting on the reverse of
the map, and alterations to the copperplate. However,
such details are beyond the scope of this work.

Speed's maps are very rarely found in 'original'
colour, the English map publishing industry being
generally less well developed at this time than the
Dutch. Most examples of Speed's maps now on the
market in colour will have been coloured within the
last century (see page 19 for further comment on map
colouring). In addition to the visual appeal and detail
on the face of the map, Speed's maps, in most editions,
have a text on the reverse describing the area shown
– this provides an amusing and fascinating insight into
that area in the early 1660s and can be displayed, if the

as a historian and genealogist, produced his maps, based
on those of Saxton and Norden with some up-dated
information, to illustrate and accompany his *History of
Great Britain*. However, the popularity of his atlas, the
Theatre of the Empire of Great Britain, was to exceed

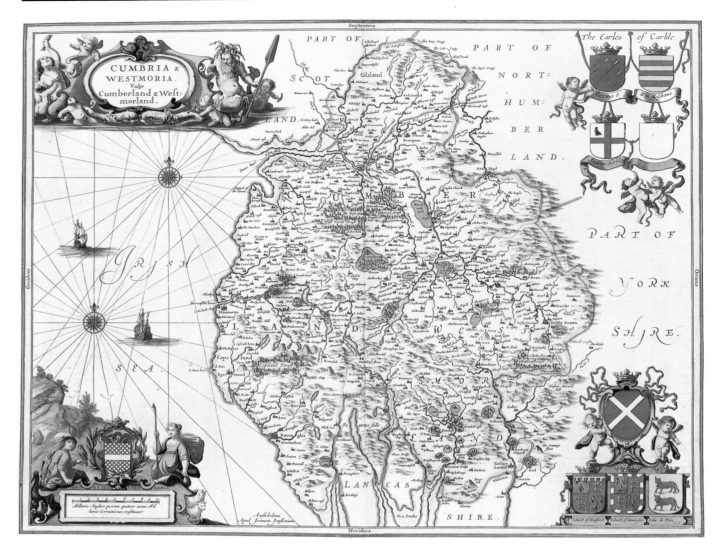

Jan Jansson's elegantly designed map of
today's Cumbria.
Amsterdam 1646
54 x 42cm 21.5 x 16.5in

*The attractive design of this map allows full scope for the
artist/engraver's skills. The figures around the titlepiece indicate
the coastal nature of the region while the scene at lower left shows
the wild mountainous terrain; sheep with a shepherdess and a
gentleman holding a falcon border the mileage scale. The coats
of arms of the county's nobility occupy the right hand corners.
Fine original colour on good clean paper (not always the case
with Jansson's maps) make this a particularly attractive example.*

map is to be framed, by having the framer place glass,
rather than a backing board, on the reverse of the
frame.

Dutch maps of the counties

During the seventeenth century Speed's work dominated English county mapping. However, comparison

with the productions of the Dutch publishing houses
of Blaeu and Jansson illustrates a dramatic contrast in
style. Although Blaeu and Jansson based their English
county atlases on the work of Saxton and Speed, their
finer artistic and stylistic approach produced maps of
elegance and beauty. The actual county area shown
and place name script is kept neat and unfussy while
lettering outside the area, particularly in Jansson's
maps, is elaborately flourished. Ships, fish, monsters
and so on occupy sea areas while the cartouches and
mileage scales allow the engraver to produce artistic,
well-designed decoration, symbolizing the nature or
produce of an area, or the occupations of its
inhabitants.

Other map makers

Blaeu's atlas was published in Amsterdam from 1645
to 1672, and Jansson's from 1646 onwards. Numerous
editions of each atlas were published, although none
had English text; the Jansson maps continued in use
through the publishers Schenk and Valk as late as 1724.

Other noteworthy map makers during the seventeenth century are: Pieter van den Keere, whose

miniature maps of the counties were published in a Dutch edition of Camden's *Britannia* (1617 and 1619), and subsequently in a pocket atlas of the English counties from 1627 – the so-called 'Miniature Speed' series of maps; Michael Drayton, whose curious, fanciful maps of parts of England and Wales, decorated with nymphs, river gods, shepherds and so on, illustrated his lengthy poem glorifying the attractions of Albion – *Poly-Olbion*; Richard Blome, who in 1673 published his version of *Britannia* and in 1681 a pocket atlas of maps, each series containing rather crudely engraved, though charming, county maps based invariably on Speed; John Seller, who produced individual county maps and miniature county atlases; Robert Morden, whose maps are particularly popular now, as then, for their clarity and detail.

Although the road maps of John Ogilby and the sea charts of Captain Greenville Collins do not fall within the category of county maps, they constitute important publications of this time, and should be given due consideration by collectors of regional county maps (see pages 84 and 85).

Into the eighteenth century

Robert Morden's work straddles 1700, a date around which the English (in particular, London) map trade was developing from the shadows of the European map makers in Holland and France and assuming an identity of its own. Around this date the following map sellers, many of whom were also map makers, can be identified: Richard Blome, Christopher Browne, Philip Lea, Robert Morden, Sutton Nicholls, John Ogilby, John, Henry and Philip Overton, John Seller, Peter Stent, Thomas Taylor and George Willdey. Others made charts or produced foreign map atlases.

Morden produced a variety of cartographic works: the folio maps, newly compiled and assembled for the 1695 edition of Camden's *Britannia* (later 1722, 1753 and 1772); a smaller series of county maps for *The New Description and State of England* (1701 and 1708) also used in *Magna Britannia et Hibernia* (1720 and 1738); a series of playing card maps (see page 189); and a large number of separately-issued maps of London, England and the rest of the World. Many other map makers of this period were equally active and English map production throughout the eighteenth century continued prolifically. Although much of the map maker's attention was drawn to foreign parts – particularly North America, India and the Far East – output of county atlases and, especially, large-scale county maps produced some attractive series of maps and a number of particularly fine cartographic publications.

As the eighteenth century progressed maps became less ornate and more practical in concept so that, at the end of the century, few atlases were still in production using the decorative, pictorial style of title cartouche typical of the mid-century. Maps by Ellis, Kitchin, Bowen, Jefferys, Gibson and many others are characteristic of the mid-century – finely engraved, detailed, produced for folio or quarto atlases, travel books or periodical magazines, and with some decoration. Herman Moll's distinctive county maps for *A Set of Fifty New and Correct Maps of England and Wales* (1724) contain a rather plain county map with, in each side margin, engravings of relics and antiquities found within that county.

The finest county atlas produced during this century was the largest scale atlas produced to date by one of the most prolific map makers of the time. Thomas Kitchin, in 1755, in association with Emanuel Bowen produced *The Large English Atlas*, each map measuring approximately 68.5 x 53 cm 27 x 21 in. There were 45 maps in all, clearly engraved in great detail with inset plans, views and numerous panels of text describing the commerce and attributes of the area, surrounding

Above
Kent

A detailed 'miniature Speed' map.
London 1627
12 x 9cm 4.75 x 3.5in

The title of this atlas explains the nick-name that this attractive series of miniature maps has acquired ... 'England Wales Scotland and Ireland Described and Abridged With ye Historic Relation of things worthy memory from a farr larger Voulume Done by John Speed ...'. Speed, in fact, had nothing to do with the publication of this atlas, but his opportunistic publisher George Humble saw the potential of acknowledging and utilizing the popular Speed's name within the title of the book and the name has subsequently stuck.

Most of the maps were engraved by Pieter van den Keere although this is not the case here – this and various other maps were copied from Speed as was the text of the volume, to replace certain engravings by Keere which had, in the Saxton tradition, grouped various counties together.

John Ogilby

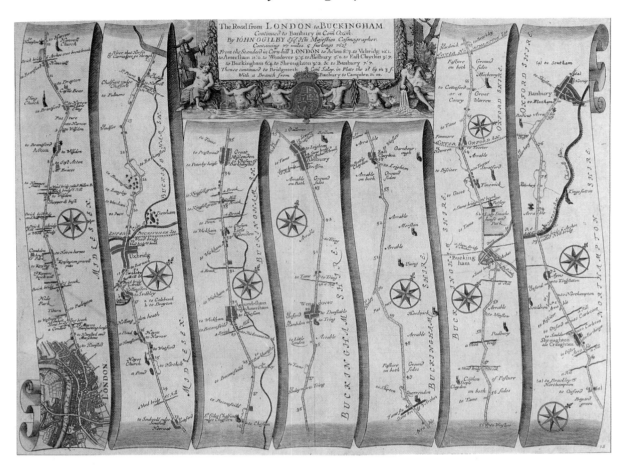

Amongst the most clearly identifiable styles of maps are the road 'strip' maps of John Ogilby. First published in 1675, the concept of the road 'unfolding' as a scroll is still followed in many route maps, each section of the road using a compass rose to identify the orientation of that particular stretch.

Ogilby's survey had been made between 1669 and 1674 and his *Britannia* was the first work to utilize solely the statute mile of 1760 yards and maintain a scale of one inch to the mile. The one hundred road maps were, in the first edition, accompanied by text describing the roads and any features of interest to the traveller. Separate map sheets from the second edition of 1675 and others of 1676 and 1698 are identified by the addition of plate numbers in the lower right hand corner.

Although there were only three editions of Ogilby's road book, there were various 'reduced' format versions published during the first half of the seventeenth century. Copies by Senex and Gardner of 1719 are frequently found as are the charming maps from John Owen and Emanuel Bowen's *Britannia Depicta or Ogilby Improved …* from 1720

'The Road from London to Buckingham

One of the famous series of one hundred 'strip' or road maps by John Ogilby.
London 1675
44 x 33.5cm 14 x 10in

to around 1764 (see page 87).

John Ogilby was a fascinating figure. Born in 1600, probably in Edinburgh, it was not until he was seventy years old that the works for which he is most famed now were published. During his life up to that time he had achieved prominence as a dancer who after an injury incurred while performing became a dancing instructor and choreographer; subsequently, under his patron, the Earl of Strafford, becoming manager and director of Dublin's first theatre. During the 1650s, after losing much of his fortune during the Civil War, he turned his attention to translating and publishing various Greek classics. This publishing venture also proved successful until the Fire of London of 1666 destroyed much of his stock. Undeterred, in the years around 1670 Ogilby set about his most ambitious publishing exercise – the production of atlases of each part of the world.

Promoting and advertising aggressively between 1669 and 1676, when he died, Ogilby saw published volumes on Africa, Japan, America, China, Asia as well as the road book of Britain. Excepting the last instance, these folio volumes were translations, almost entirely of contemporary works by Dutch and German authors and were well illustrated with engravings and maps, also invariably copied from the original publications. Besides the entirely original road book, one of the most important British cartographic developments of the period – Ogilby added new information to the original text of Arnoldus Montanus' book on America, producing one of the best records of colonial America of the time. Ogilby's intended volume on Europe never materialized – conceivably since there was no model on which he could produce such a work – but the volumes already produced ensured Ogilby died a rich man.

Most county maps after the publication of *Britannia* included roads testifying to the acceptance and high regard in which Ogilby's maps were held.

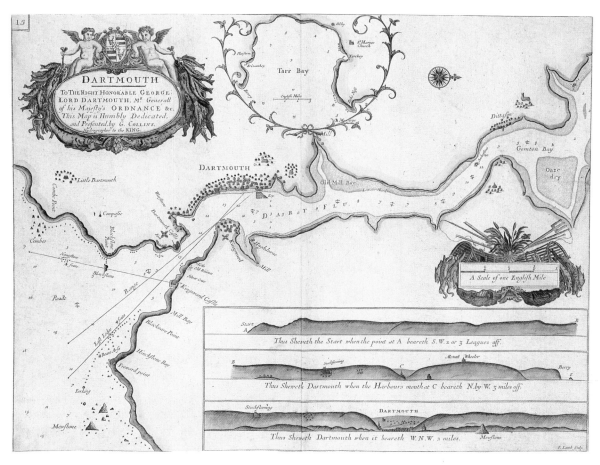

Above
Dartmouth

A fine example of Collins' charts, here
with an inset of Torbay.
London 1693
57 x 44.5cm 22.5 x 17.5in

*The particular importance of Dartmouth
as a fishing port is indicated by the bountiful
catch displayed around the title cartouche.*

In 1693, *Great Britain's Coasting Pilot*
was published in London with charts
by Captain Greenville Collins. As a
developing maritime nation, England's
own printed chart production was negligible compared with the Dutch. John
Seller's 1671 *English Pilot* for the greater
part had merely reprinted out-of-date
Dutch copperplates and in 1681 Samuel
Pepys, Secretary of the Admiralty,
Master of Trinity House, and himself a
great collector of maps, appointed
Collins to survey the coasts of England,
Scotland, Wales and the many adjacent
islands.

After several years' surveying, Collins
employed a number of London's leading
engravers including Moll, Harris, Lamb
and Moxon to cut the plates which were
finally published some twelve years after
the original commission. Although
there were considerable inaccuracies,
this survey was a great achievement in
the history of British hydrography.

During the next century Collins' atlas
was republished about twenty times
(without revision of the original charts)
by the firm of Mount and Page.

Left
Great Britain's Coasting Pylot

The title page to Greenville Collins'
atlas of sea charts.
London 1693-1723
26 x 41cm 10 x 16in

*This fine engraving embodies the spirit of
British maritime ambitions at the end of the
seventeenth century. Britannia and Neptune flank the title, held aloft by a merman.
The Royal Coat of Arms is draped by
fanfare-blowing cherubs and, below, a
chart of the British Isles is held by a merman
and a mermaid. Sea nymphs at each lower
corner display navigating instruments including a plumb line and a cross staff.*

*This is the second state of this engraving
– the first state had the date 1693 in the
'neck' of the shell. The date was removed
for the second edition and in 1738 the
Hanoverian arms replaced those of the
Tudors.*

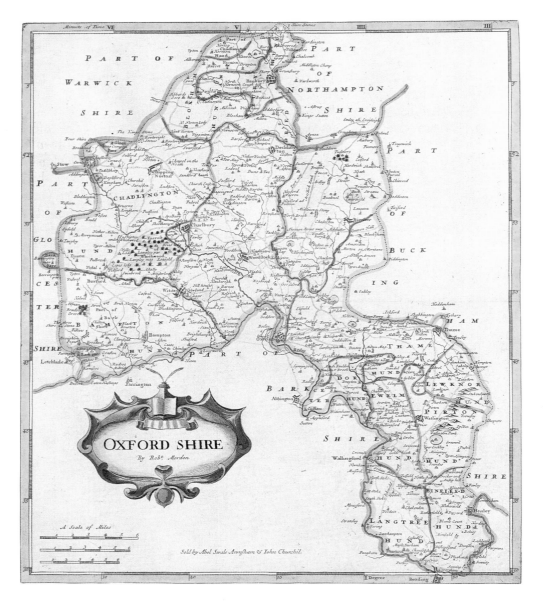

Left
Oxford Shire

Robert Morden's detailed map of the county.
London 1695
36 x 42cm 14 x 16.5in

Robert Morden's maps were prepared for and published in Edmund Gibson's revised edition of William Camden's 'Britannia' in 1695. This popular history and topographical guide book was issued in further editions through the eighteenth century, but the maps are sought by collectors now for their attractively simple design and clear style, their detail, and their reasonable price in comparison to other seventeenth-century county maps.

each county shown. This atlas was published about a dozen times over the next 40 years, a period during which large-scale county mapping and the discipline of precise surveying techniques laid the foundation for the Ordnance Survey of the next century.

Large-scale maps of all counties and many towns can be found from this period – many at scales as large as over one or two inches to the mile. Amongst a variety of map makers the names of the following are predominant: Rocque – maps of Surrey (1768), Middlesex (1754), Berkshire (1761), and Shropshire (1752); Jefferys – maps of Bedfordshire (1765), Buckinghamshire (1770), Huntingdon (1768), Oxfordshire (1768–9), Staffordshire (1747) and Yorkshire (1771); Taylor – maps of Dorset (1765), Gloucester (1777), Hampshire (1759), Hereford (1754) and Worcestershire (1772); in addition, the following produced notable maps: Gascoigne and Martyn – Cornwall (1700 and 1748 respectively); Donne–Devon (1765); Chapman and André – Essex (1777); Dury

and Andrews – Hertfordshire (1766); Andrews, Dury and Herbert – Kent (1769); Davis of Lewknor – Oxfordshire (1797); Budgen and Yeakill/Gardner – Sussex (1724 and c 1780 respectively); Warburton – Yorkshire (1720).

In 1787, John Cary published his *New and Correct English Atlas*. Although they lack decoration, the maps have great character and are finely engraved and detailed. The popularity of this quarto atlas ensured its success and editions appeared as late as 1831. A pocket atlas entitled *Cary's Travellers' Companion* was first published in 1789 and a larger-scale folio atlas, *Cary's New English Atlas* appeared in 1809. Cary, rightly acknowledged as one of the leading map makers of his day, also issued world atlases and globes.

The nineteenth century

In the first half of the nineteenth century, the only 'decoration' map makers allowed themselves was, perhaps, the addition of a vignette scene of a city, church or cathedral, or country seat. Fullerton, Langley, Pigot and the Greenwoods, amongst others, all used this format. The latter pair of brothers, Christopher and James, also produced some fine large-scale county maps. The famous exception to this rather austere approach was, of course, Thomas Moule who, with his intricately decorated maps, produced in 1836 the collection which is commonly known as 'the last series of decorative county maps'.

Production of 'county' atlases after about 1850 began to decline rapidly in the face of competition from the Ordnance Survey and its 'sectional' maps relegating the importance of county boundaries. The Ordnance Survey, administered by the Board of Ordnance, was set up in 1791 to satisfy the military need for accurate maps of the country at a time when invasion by France was a distinct danger. The first sheets, of Kent, were completed in 1801 but it was 70 years before the rest of the country was mapped at the uniform scale of one inch to a mile.

In addition to the mainstream of county maps, mention has already been made of Michael Drayton's maps and of playing card maps of the seventeenth century. Also worthy of note are George Bickham's 'bird's eye view' maps of the counties, first published in 1754, and a variety of miniature maps of the early nineteenth century, which are treasured by collectors for their curious presentations. Amongst these are the circular maps of John Luffman (1803), the charming decorated maps by Aristide Michel Perrot (1823) and the children's atlas maps by Reuben Ramble (1845).

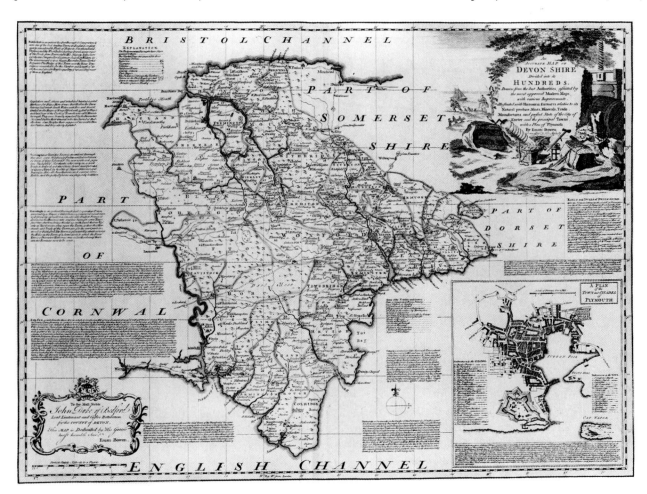

Above
Devon Shire

Emanuel Bowen's detailed map of Devonshire.
London c 1755
69 x 53cm 27 x 21in

The 'Large English Atlas' with maps by Emanuel Bowen and Thomas Kitchin is justifiably regarded as one of the best county atlases published to that date. Combining cartographic detail with descriptive annotations, a vast amount of information was shown in these engravings. The large scale of these maps had not previously been attempted within an atlas although some of the counties were being mapped at larger scale.

Descriptive notes around the map area concerned the history, agriculture, economy, natural features, and people of note within that county and most of the maps included either plans, or panoramas, of the major towns within the county.

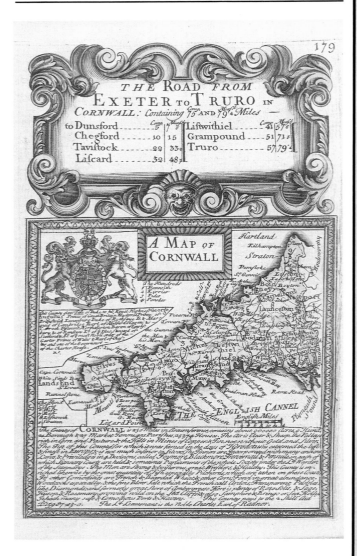

Above
A Map of Cornwall

John Owen and Emanuel Bowen's
attractive, detailed map of Cornwall.
London 1720
Map size 12 x 12cm 4.7 x 4.7in

*This popular and attractive series of maps appeared in one of the
many 'road' books which followed in the wake of John Ogilby's
'Britannia'. This particular volume 'Britannia Depicta … or …
Ogilby Improved' has maps of each English and Welsh county
and numerous 'strip' maps of the major post roads. Each county
map is topped by a distance chart and the surround of the county
is closely packed with detail concerning that county – its products,
economy, agriculture, market days and so on.*

The English Counties Delineated

Thomas Moule's maps are probably, with those by Speed,
the best known of all series of English county maps.

Issued from 1836, the maps combine a clarity of
cartographic style with immense detail by way of vignette
views, scenes and portraits relating to the county shown,
often set within a gothic architectural or floral surround into
which armorial devices and so on are worked. As the editions
of the maps were published, first in *The English Counties
Delineated* and later in *Barclay's English Dictionary*,

Right
Berkshire

A typically decorative map
by Thomas Moule.
London c1850
26 x 20cm 10 x 8in

*In this map we see a fine
example of Moule's combination of cartography with informative decoration. Armoured
knights flank the county which
is adorned, above, by the
royal coat of arms. Circular
views of Virginia Water,
Windsor Castle, the Town
and Bridge of Windsor, and
Abingdon town hall appear in
each corner and two elegant
maidens symbolize the tranquil
and genteel nature of the River
Thames and its countryside.
The county's hundreds, roads
and railways are shown,
as are parklands and relief
features such as the ridge of
hills which runs across the
county east to west.*

the development of the network of railways throughout England can be observed.

Moule, like many other map makers and map sellers before him, was a man of many talents. As an author his output included books and papers on topography, history, genealogy, heraldry and architecture, and the maps which he designed show elements of each of these studies. The 57 maps and plans produced for Moule's *English Counties …,* originally issued as a 'part-work', include maps of each English county, the towns of London, Bath, Boston,

Portsmouth, Plymouth and the Isles of Wight, Man and Thanet.

Early issues of these maps may sometimes be found in contemporary colourwash but the majority of examples seen on the market now will be in relatively recently applied colour. Frequently titled 'the last decorative series of county maps' these are good, informative maps which are as popular with collectors today as they were when published during the early years of Victoria's reign.

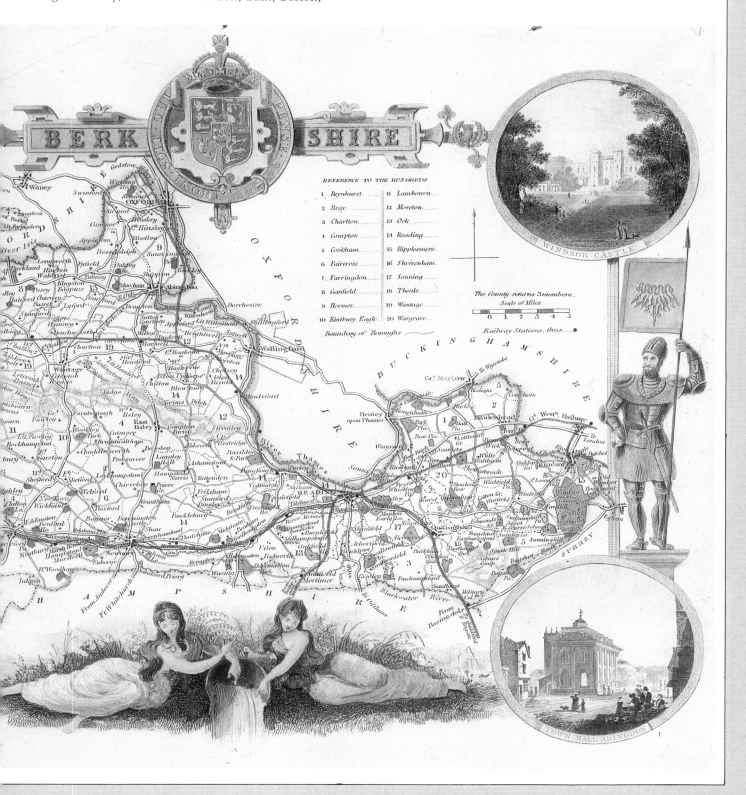

BRITISH TOWN PLANS

The earlier chapter on plans issued in general atlases of a world-wide content should also be read in relation, particularly, to the publication of maps showing London and a few other major British cities and ports.

Of all the world's cities none has been shown on printed maps more often than London. Appearing as the first plate in Braun and Hogenberg's *Civitates* of 1572, it has remained in all other multi-national collections of town plans printed in Europe and, obviously, in England. As such, a collection of London maps alone would involve hundreds of different publications regardless of variant editions of one map.

London maps

London maps, like many other city plans, may be of a small scale showing the whole town or sectional, in large scale, showing much greater detail. In the case of London, there are also the various series of Ward plans of parts of the City which were engraved during the eighteenth century to adorn such works as John Strype's edition of Stowe's *A survey of the Cities of London and Westminster* (1720), William Maitland's *History and Survey of London* (1756) and others.

The Braun and Hogenberg map, about whose origins very little is known, was the first printed plan of London. In later editions of the same work were included plans of Cambridge, Oxford/Windsor (on one

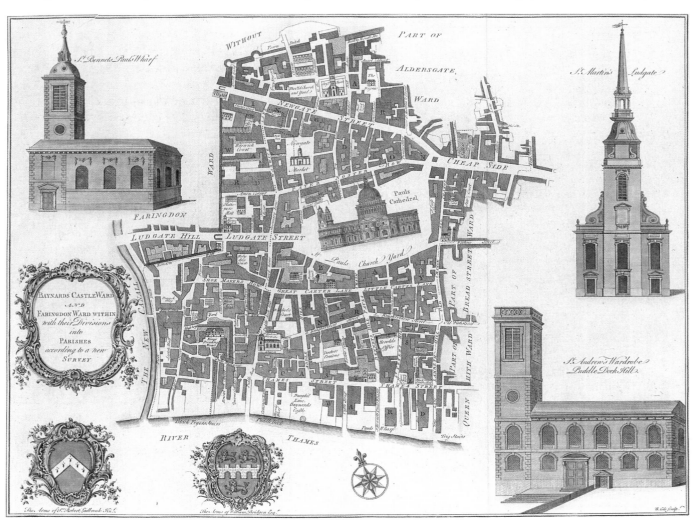

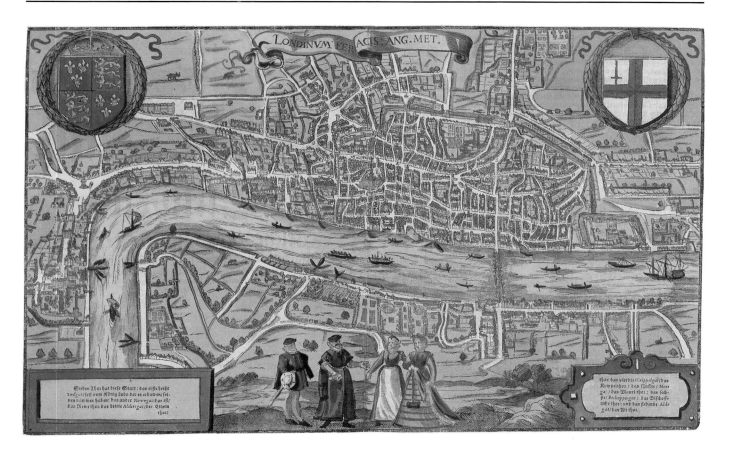

Above
Londinum Feraccis Ang. Met.

Sebastian Munster's *Cosmographia* was
issued, after his death in 1552, by his
son-in-law Henrici Petri who later
incorporated town plans into the work.
Basle 1598
35.5 x 22cm 14 x 8.5in

*The great town plans book 'Civitates Orbis Terrarum' of Braun
and Hogenberg had, of course, included a plan of London. Petri,
for Munster, copied many of the 'Civitates' originals for inclusion
in the 'Cosmographia', and this woodblock, although slightly
reduced in size from the original is a faithful copy.*

*The map shows the Cities of London and Westminster and
describes clearly the extent of early Elizabethan London. The
Tower of London and the City Wall can be seen, as can Saint
Paul's and London Bridge.*

Left
**Baynards Castle Ward and Faringdon
Ward Within**

Benjamin Cole's detailed Ward plan from
William Maitland's *History and Survey of
London.*
London 1756
46 x 36cm 18 x 14in

*A decorative combination of detailed plan and architectural
profile. Of similar scale and detail are the plans from Stowe's
1720 survey. Later miniature versions by Noorthouck, Harrison
and Thornton can also be found.*

plate) (1575); Norwich, Bristol, Chester and Edinburgh (1581); Canterbury (1588); The Palace of
Nonsuch (1598); Exeter, York/Shrewsbury/Lancaster/
Richmond-upon-Thames (on one plate), Dublin/Gal-
way/Limerick/Cork (on one plate) (1617). As we have
seen many of these were also copied by Valegio and
Meisner.

In 1611 John Speed's *Theatre of the Empire of Great
Britain* was published, with most of the county maps
having inset plans of the major towns within that area.
Speed used the existing material where possible,
including maps by William Smith, and for London
and Westminster on the Middlesex map, those by John
Norden; where necessary, if surveys were not available,
Speed himself visited and plotted the town. Speed's
atlas was particularly popular and ran into many
editions and the town plans, also, were well regarded.
These were copied as individual maps by Rutger
Hermannides in 1661, and Johann Christian Bear in
1693, and incorporated onto many other map makers'
county maps.

During the seventeenth century few provincial towns
had been surveyed, and in the case of London, the
greatest surveying activity, was, in fact, precipitated
by disaster – the Great Fire of 1666 – and a large
number of both home-produced and foreign surveys
were printed to record the extent of the fire and its
effect. Several of these maps were published in
'broadsheet' form and, as such, are particularly rare.

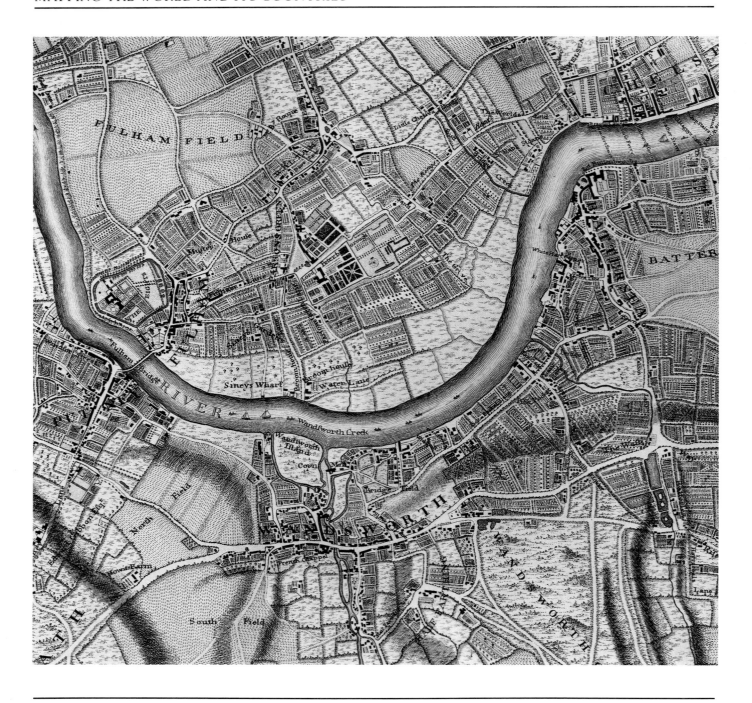

Above

A section from John Rocque's detailed map of London and its environs – covering Fulham, Wandsworth, Chelsea and Battersea.

London 1746
Original sheet size 61 x 48cm 24 x 19in

Rocque's map of London provides, like many of his other large-scale plans, a fascinating record of the relationship of town and country in the mid-eighteenth century.

John Rocque was a Huguenot emigré surveyor who worked in London from about 1734. He produced numerous large-scale surveys of estates, counties and towns, county atlases and foreign and national maps. However, his best known works are the two surveys which he compiled of the Cities of London and Westminster, at a scale of nearly 25 inches to the mile and

identifying each building, and, his map of the country around London – done on a scale of about 5.5 inches to the mile. The small section of the map illustrated here gives an impression of the detail which Rocque included and reflects his early training as 'dessinateur de jardins'.

The section shown is known to many Londoners today and it is still clearly possible to identify road patterns which have been maintained since this period – the King's Road, Parsons Green and many other areas are familiar to us now. In the mid-eighteenth century the area was becoming important to London and Westminster for its market gardens, clearly seen here, and the small villages on the roads out of London were gradually becoming enmeshed in the expanding metropolis. However, it was another one hundred years or so before the area became extensively built up.

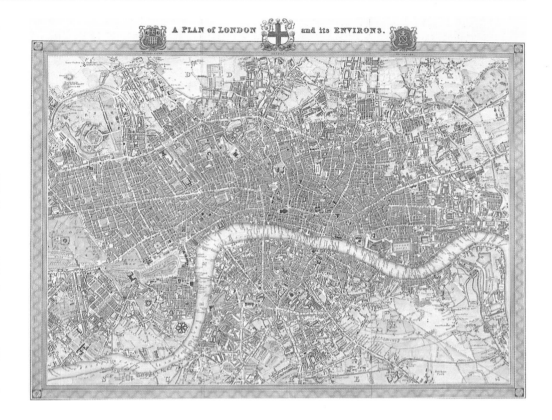

A PLAN of LONDON and its ENVIRONS.

Right
A Plan of London and its Environs

A detailed, attractive, map issued in Samuel Lewis's *Topographical Dictionary*.
London 1831-48
46 x 33cm 18 x 13in

This detailed map shows the main extent of London around 150 years ago. Development to the west, not shown here, incorporated Kensington, Chelsea and Paddington. There were, inevitably, a great number of maps of London published during the nineteenth century and the growth of the town can be clearly seen. However, many of those maps were issued in dissected, mounted and folded form in slip cases.

One of the most prolific engravers working in London at this point was Wenceslaus Hollar. Born in Prague in 1607 he came to England in 1636 and produced engravings on innumerable subjects, including some fine panoramas of London. Plans of Hull, Oxford and Dover are known and some eight London plans are attributed to him.

With the exception of London maps, the majority of town plans being issued still appeared in books, possibly county history books or guide books, and it was during the eighteenth century, as the techniques of large-scale mapping progressed, that new detailed plans of British towns appeared. In the forefront of new style was the immigrant Hugenot, John Rocque.

There are few books of British town plans although some of the World atlases, which incorporated plans, have a predominance of British towns. Listed below are some of the publications in which plans can be found.

1764 – A Collection of Plans of the Principal Cities of Great Britain and Ireland... Maps reduced from surveys of John Rocque, engraved by Andrew Dury. Attractive miniature maps, usually found coloured. London, Kensington, Oxford, Lewes, Exeter, Chichester, Bristol, Bath, Bury St Edmunds, Shrewsbury, York, Chester, Boston, Aberdeen, Edinburgh, Dublin and Cork.

1771 Collection of Plans of the Capital Cities of Europe... by John Andrews. London, Edinburgh and Dublin.

1810 The British Atlas... A county atlas with town plans by John Roper and John Cole, published in London. Includes Bedford, Cambridge, Canterbury, Carlisle, Chester, Colchester, Coventry, Derby, Durham, Exeter, Gloucester, Hereford, Liverpool, Manchester, Newcastle, Northampton, Norwich, Oxford, St Albans, Winchester, Worcester and Newport (inset on map of the Isle of Wight).

1837 The English Counties Delineated... by Thomas Moule. Includes plans of London, Cambridge, Bath, Oxford and Boston.

c 1840 Maps for the Society for the Diffusion of Useful Knowledge Published initially by Baldwin and Craddock. Included plans of London, Edinburgh, Dublin, Birmingham and Liverpool.

1851 The Illustrated Atlas of the World by John Tallis. This interesting atlas includes finely steelplate-engraved plans of the cities listed below. All the maps and plans have ornamental borders and most have vignette scenes; they are invariably described as 'the last decorative series of maps....': Bath, Birmingham, Bradford, Brighton, Bristol and Clifton, Exeter, Leeds, London, Liverpool, Manchester, Newcastle, Plymouth, Devonport, Preston, Sheffield, Southampton, York, Aberdeen, Edinburgh, Glasgow, Perth, Belfast, Cork, Dublin.

SCOTLAND AND IRELAND

Neither Scotland nor Ireland enjoyed the attention of Christopher Saxton's magnificent survey of 1579. Accordingly, with little original material to work on the atlases of the early seventeenth century showed little detailed mapping of these countries.

The printed cartographical history of these areas commences, invariably, with the earliest issue of Ptolemy's *Geographia*. The outlines of both countries are poor, Scotland being notable for its eastwards slant, while Ireland is shown too small and too far north.

Cartographic development during the sixteenth century is seen on those maps of the British Isles already described, from which the first separate maps of each country were developed. Many of these separate maps, engraved and published in Italy, were based on George Lily's map of The British Isles although the long-established trading links between Italy and Ireland led to the incorporation of detail from contemporary portolan charts. In 1564 Gerard Mercator's influential map of the British Isles was published and it was on this that Ortelius, in 1573, based his maps of Scotland and Ireland, the first generally available separate maps of those two countries. Mercator's 'Atlas' of 1595 also included separate maps of each country and from this date on it is not difficult to find maps of each country published in England, the Low Countries and France. However, the regional mapping of each of these two nations does not follow the same prolific path as that of England and Wales.

Above
The Kingdome of Scotland

The early edition of John Speed's famous map.
London 1610-11
51 x 38.5cm 20 x 15in

This attractive map was published in the 'Theatre of The Empire of Great Britain' from 1610–11 but in 1652, during the Interregnum, the plate was re-engraved, replacing the Royal figures shown here with more lowly Scottish characters.
This plate was engraved by Jodocus Hondius in Amsterdam, as were others in the atlas, and based on Gerard Mercator's map of the country published in 1595. As the copperplate ages the 'silky' effect engraved in the sea areas fades and there is a noticeable premium on the price of a good early issue of this map.

Left
Scotiae Tabula

Abraham Ortelius' attractive map of Scotland –
the first generally obtainable of the country alone.
Antwerp 1573
47.5 x 35.5cm 18.5 x 14in

A handsome map used in all editions after 1573. Orientated with north to the right, this map is in good original colour, and is distinguished by the large framed titlepiece. This map, like that of the British Isles, was based on Mercator's 1564 map of the British Isles utilizing the northern outlines.

Scotland

New regional maps of Scotland are found in the editions of the Mercator/Hondius 'Atlas' from 1630, entitled *A New Description of the Shyres Lothian and Linlitquo*, and *Orcadun et Schetlandiae* in 1636. These maps were both supplied by Timothy Pont, the most famous name in seventeenth-century Scottish map making.

Timothy Pont was a minister, born about 1565. Travelling widely throughout Scotland, surveying the countryside, he compiled a collection of manuscripts which, some 50 years after being draughted, and having been redrawn by Robert Gordon, found their way to Amsterdam. Here, in 1654, they were engraved and published, with text by Sir John Scot, by W and J Blaeu. There are 46 maps of Scottish counties or regions, of which the majority are attributed to Pont who had died in 1610, unaware that his work would actually be published, and would remain the largest scale series of Scottish maps until the nineteenth century.

During the eighteenth century, there are several maps which can be identified as distinct improvements on those previously available. These include: Herman Moll's large map of the *North Part of Britain called Scotland* (1714), embellished with views of the major

95

cities along each side; James Dorret's map of 1750; and John Ainslie's map of 1789. Throughout this time versions of these maps appeared in general atlases and reduced in county map format in the atlases of Moll (a set of *Thirty-six New and Correct Maps of Scotland*,

Below
Irlandiae

Baptist Boazio's beautiful map from Ortelius's atlas.
Antwerp 1602
54 x 42.5cm 21.5 x 17in

Boazio's map had been drawn and published in London, about three or four years prior to this atlas's publication. In 1601 Johann Baptist Vrients took over publication of the 'Theatrum Orbis Terrarum' and incorporated new maps, amongst which was this one by Boazio.

The example illustrated is in good condition and fine original colour. Unfortunately, the large copperplate size of this map necessitated either trimming the sheet very close when binding, or folding the edges of the map in. Consequently, the map is not often found in good condition. Published in relatively few editions of the atlas, this is one of the most desirable maps of the island for collectors.

1725), Kitchin (*Geographica Scotiae*, 1749) and Mostyn John Armstrong (*A Scotch Atlas*, 1777). Towards the latter part of the century, larger-scale, separately-issued county maps appeared as did road strip maps in a number of atlases, the most commonly found being those from George Taylor and Andrew Skinner's *Survey and Maps of the Roads of North Britain or Scotland*, published in 1776.

As with many other areas during the eighteenth century, the military requirement for detailed maps following the Jacobite uprising of 1745, was an important factor in the cartographic development of the area. Commenced in 1747, under the direction of William Roy, a survey of the Highlands was completed in 1755 and, although never published in its own right, it was incorporated into Aaron Arrowsmith's excellent and influential map of 1807. William Roy himself, eventually promoted to Major General, became one of the founders of the Ordnance Survey, his work in Scotland having set the standard for much subsequent surveying and mapping by British officers in North America, India, Ireland and finally the remainder of Britain.

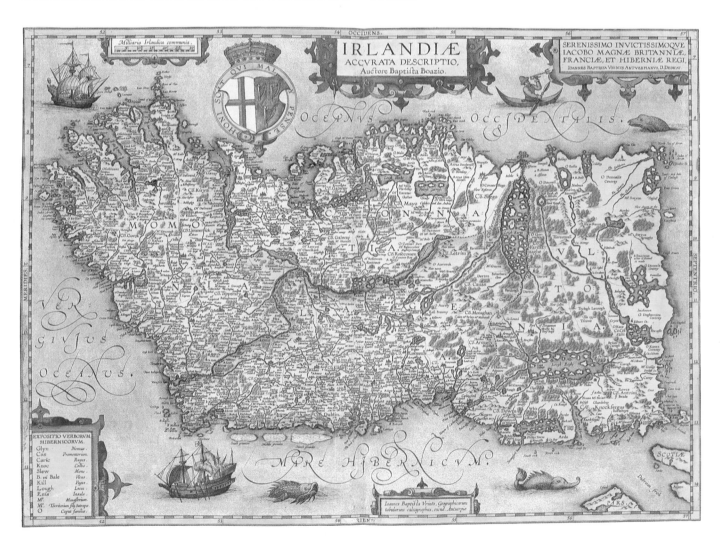

An uncommon series of plans which should be noted is that in *Wood's Town Atlas* of 1828, 48 plans show 47 towns surveyed by John Wood between 1818 and 1826. In many instances these are the earliest plans of those particular towns. Edinburgh had been mapped in George Braun and Frans Hogenberg's book of plans in 1581, but very few other Scottish towns had appeared in detail until this time.

A further interesting collection of county maps is found in John Thomson's *Atlas of Scotland*, published in Edinburgh in 1832. Fifty-eight map sheets cover the counties and islands on scales not previously published in atlas form – larger counties occupy two or four sheets – and not superseded until the publication of the Ordnance Survey sheets. This is a fine atlas, with which there is no comparison in British county map production, and all the more remarkable considering the great financial problems encountered by the Thomson Company due to the cost of the book's preparation and production.

Ireland

Abraham Ortelius's map of Ireland, first published in 1573, was replaced in the 'Atlas' in 1602, by a magnificent map by Baptista Boazio (see opposite). However, the most influential early seventeenth-century map was that published by John Speed in Book IV of his British Atlas of 1611. There are maps of the whole country, with border panels depicting six figures in national costume, and four maps of each of the provinces – Ulster, Leinster, Connaught and Munster – each having inset plans of the major towns within that area. These regional maps were copied by Blaeu and Jansson, and the general map by numerous others in England and Europe.

As in the case of Scotland, however, the most important cartographic event of the century was the publication, in 1685 of Sir William Petty's *Hibernia Delineato*, the first atlas of the country. This comprised thirty-one county maps and five others, and the information therein was quickly adopted by subsequent map makers.

Herman Moll in 1728 and Bernard Scalé in 1776 each produced small-scale county atlases, and a year later Taylor and Skinner produced a road atlas. Additionally, there were a great number of single sheet maps, many at good scales, produced in atlases or as separate wall maps during this period. A fine example is the four-sheet map by John Rocque (see page 92) of 1765, which was subsequently reissued in the Laurie and Whittle series of atlases. A notable character publishing maps, charts and atlases during the middle of the century was George Grierson, a Dublin printer and publisher who not only reissued Petty's atlas but also reissued atlases by Moll and chart books by Mount and Page, merely superimposing his own imprint.

During the nineteenth century, various minor county atlases were issued but the most important cartographic development was that of the Ordnance Survey, which was completed between 1829 and 1858. Unlike Scotland, there was relatively little home-produced map work and so the range of maps available to the collector is not, unfortunately, as great.

EUROPE

Within such a broad heading it is not possible, given the scope of this work, to describe in great detail the cartographic history of each European country. Nevertheless, the following notes should provide an outline and describe some of the map makers whose work is available to collectors.

With Europe as the centre of production of this study of printed maps, there is inevitably a vast range of European material available produced throughout the period of printed cartography. Earlier chapters have indicated how various countries became in turn the centres for map production, i.e., Italy, the Low Countries, France and England, and, at various times throughout, Germany.

At the time around 1500 when both Italy and Southern Germany were producing printed maps, Spain

Below
Tabula Nova Prima Europa

Sebastian Munster's woodcut map of Europe.
Basle 1540
33 x 25cm 13 x 10in

A boldly engraved woodcut map, orientated with north to the foot of the page, issued in both the 'Geographia' and the 'Cosmographia'.

and Portugal were Europe's leaders in marine exploration and consequently chart making. One might assume that this pre-eminence would result in a successful map publishing industry, but this was not the case. The manuscript charts which were produced were highly prized documents of great national importance and were jealously guarded. Consequently, their numbers were strictly controlled and printing copies for mass circulation was unthinkable. Such national security interests applied to all countries with maritime interests as far as charts were concerned. Similarly, a hundred years later, the Dutch East India Company exercised tight control over their manuscripts, though by this time the advances in printing techniques and the increased demand for charts necessitated wider publication within relatively short periods of time.

Maps of each southern European area (Portugal and Spain being treated as one) appeared from the first edition of *Geographia* in 1477. Subsequent issues of *Geographia* gradually incorporated 'new' maps to match all the old Ptolemaic ones. In 1482, in the Ulm edition (the first printed north of the Alps), the first map of northern Europe, i.e., Scandinavia, appeared.

Gastaldi's 1548 issue of *Geographia* added further maps of parts of Europe and, from 1570, Ortelius's atlas

Above
Europa

The earliest state of Guillaume Blaeu's magnificent map of Europe – before his change of name.
Amsterdam 1617
58 x 41cm 23 x 16in

This being the earliest known state of this map, close study of the engraving shows just how fine the quality of Blaeu's work was. Although later issues of his maps still look good it is only from the very earliest impressions that the quality of engraving is apparent.

Blaeu engraved this, and other maps of the continents and the world, around 1617-18 but the maps were not published in atlas form until his own atlas, 'Appendix' (1630) was published. By this point Blaeu had changed his name from Janssonius to the family nickname by which he is now known.

The map is exceptionally detailed, as are the town plans, copied from those in Braun and Hogenberg's 'Civitates...'. Unlike most of Blaeu's maps this is known in four different states ...

First state: dated 1617, signed Janssonio.
Second state: date removed.
Third state: signature altered to Blaeu.
Fourth state: the mythical island of Frisland (south-west of Iceland) has been removed from the copperplate.

included maps of European countries and detailed maps of many of those countries' provinces. Mercator's maps subsequently combined with those of Hondius and Jansson, and the maps of Blaeu's *Atlas Major* covered in detail each part of Europe, so that by about 1650, each of the two rival firms was publishing around 400

maps of individual provinces or countries of Europe. In addition, from around 1500, maps and plans were published separately, or in history and travel books – in volumes such as the *Nuremberg Chronicle* (1493), Sebastian Munster's *Cosmographia* from 1544, Johann Stumpf's *Chronik* of 1548 and, of course, those maps issued in Italy under the general heading of 'Lafreri' maps.

With Braun and Hogenberg's town plans and the numerous atlases and books with maps issued as early as 1650 or 1660, there is great scope for the collector for maps of the whole country or just one region.

I list below by country, in order of national volume of output, some of the major map makers and events of relevance to European cartography.

The Low Countries

The Golden Age of Cartography would be identified as the greater part of the seventeenth century with particular reference to those maps and atlases produced in Amsterdam at this time.

The Dutch, as a mercantile nation, having broken the yoke of Iberian domination overtook Spain and Portugal as Europe's leading traders, and much of the

Below
Nove Europae ...

A decorative map from the combined map-publishing talents of Jodocus Hondius and Jan Jansson.
Amsterdam 1623-1638
55 x 41cm 21.5 x 16in

A decorative, uncommon carte à figures which appeared in only a few editions of the atlases produced by the collaboration of Hondius and Jansson. Engraved by Hondius in 1623, these maps did not appear in atlases until 1631. However, the large size of the engraved map made binding without trimming the edges difficult and so the publishers preferred to use other smaller, less decorative continent maps.

Views of Lisbon, Toledo, London, Paris, Rome and Venice appear along the top border, while panels along each side show English, French, Belgian, Castilian, Venetian, German, Hungarian, Bohemian, Polish and Greek figures.

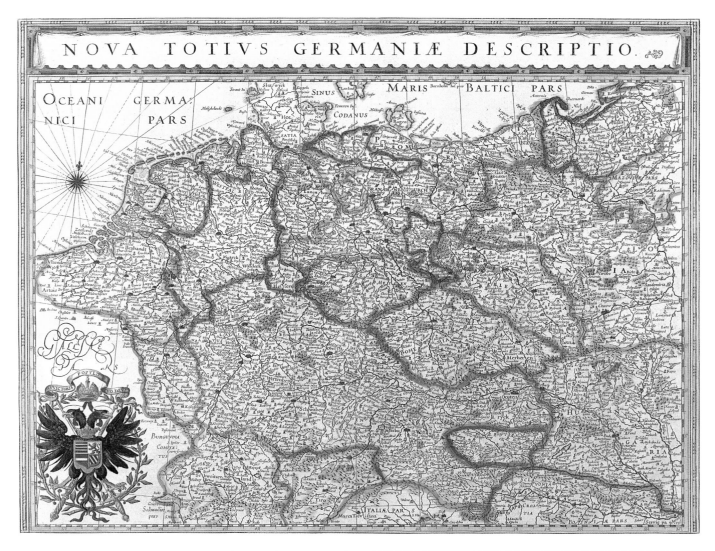

NOVA TOTIVS GERMANIÆ DESCRIPTIO.

Above
Nova Totius Germaniae Descroptio

Joan Blaeu's finely engraved maps of
Germany.
Amsterdam c1635
50.5 x 40cm 20 x 16in

A detailed map of Germany including also the Low Countries, and much of Central and Eastern Europe. The title is effectively, though unusually, shown here on a banner extending the whole width of the engraving.

wealth of the eastern and western hemispheres passed through the Low Countries. Inevitably, Holland's role as a merchant naval power necessitated an active chart and map making industry and from the mid-1500s the works of among others, Mercator, Ortelius, and de Jode, laid the foundations for the greatest period of map production the world had seen, arguably the greatest of all periods.

Other chapters describe the maps accompanying Linschoten's book of voyages (see page120) which, in both geographical content and decorative style, were to become so influential. During this period, around 1600, some magnificent maps were drawn and engraved by Petrus Plancius, Baptist and Jan van Doetechum for amongst others, Waghenaer and Pieter van den Keere. However, the emphasis here was on maps of the World, the Indies and routes to them, as opposed to European provinces.

Throughout this period Ortelius's and Mercator's atlases were being reissued and in 1630 W J Blaeu issued his *Atlantis Appendix*, the first of his sequence of World atlases which were to reach their climax in the *Atlas Maior* of 1662. Amsterdam at this time supported a vast number of artists, map makers, engravers, printers and publishers, of whom some of the more prominent figures are listed below.

The work of Joan Blaeu (see page 41), Jan Jansson (see page 39), Pieter Goos (see page 61) and the Van Keulens (see page 62) is described elsewhere.

Atlases of sea charts were issued by Jacob (father) and Arnold (son) Colom; by Anthonie (father), Jacob (son) and Caspar (son) Jacobsz, the two sons trading under the name 'Lootsman' (translation of sea-pilot) to avoid confusion with another printer then active in

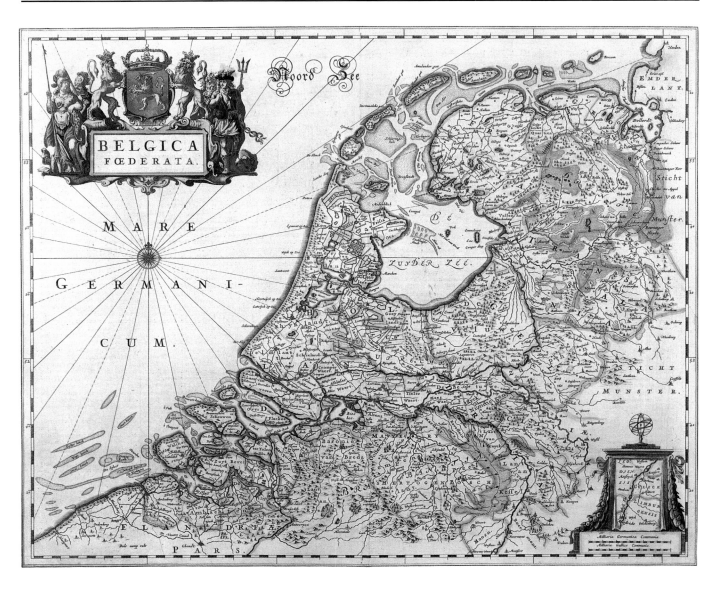

Above

Belgica Foederata

Joan Blaeu's finely engraved map of the newly independent provinces of the Netherlands.
Amsterdam 1662
53 x 44cm 21 x 17.5in

A finely engraved map, with large elegant title cartouche, in contemporary colour from the magnificent 'Atlas Major'.

Amsterdam. Sea atlases and charts were also produced by the father and son business of the Hendrick Donckers; Pieter van Alphen; Arent Roggeveen; Jacobus Robijn; and Frederick de Wit (whose charts were later issued by Renard). With sea atlases from Blaeu, Jansson, Goos and the Van Keulens, output was enormous during this century, but charts, in good condition, by many of these figures are hard to find. Land, as opposed to marine, atlases were also produced in great numbers. In addition to those of Blaeu and Jansson, fine maps can be found by the Visscher family; Frederic de Wit; Hugh (father), Carel (son) and Abraham (grandson) Allard; Cornelis (father), Justus (son), Theodorus and Cornelis (grandsons) Dankerts; the Pieter Schenks of three generations; Gerard (father) and Leonard (son) Valck; and Johannes de Ram.

Between 1700 and 1730–40, the work of many of the above map makers continued to be published, especially by the firms of the Ottens family, and of Covens and Mortier. Pierre Mortier, of French extraction, had established a publishing house in Amsterdam by about 1685 and, utilizing an excellent business relationship which he had fostered with Parisian booksellers, set about publishing and reissuing French books including, in addition to Dutch atlases, the work of Sanson and Jaillot and, subsequently, of de L'Isle. After his death in 1711 his family continued the business, going into partnership with Johannes Covens from 1721.

Another Frenchman whose work was published in Amsterdam was Henri Abraham Châtelain whose maps for the *Atlas Historique* include one of the most spectacular maps of the New World ever published (see page 139).

Amsterdam's reign as the foremost map publishing town was really over by about 1740, but two other figures of the period should be mentioned. Between about 1710 and 1730 Pieter Van der Aa in Leiden produced in vast quantities reissues and re-engravings of many of the map and view plates which had been engraved in the previous century. Little of his output was original although that which is has a distinct style of its own (see pages 56 and 107), and is particularly sought after nowadays. Isaac Tirion's atlases of around 1750 included some good maps which were often also inserted into travel and history books of the time.

Just as Tirion's maps were based on de L'Isle's work so the maps, which are sometimes seen, engraved by Van Tagen and issued by Bachienne (c 1780), were based on those by Emanuel Bowen or Jacques Nicolas Bellin. Maps encountered with the imprint of J B Elwe published after 1790, were reissues of earlier plates (some from the previous century) and have the somewhat dubious distinction of including some of the most out-of-date geographical information seriously published; hence Elwe's map of the World is probably the last published showing California as an island! The map, however, is quite decorative.

A relatively late, but by no means less interesting, publication of note, is Philippe Vandermaelen's *Atlas Universal* published in Brussels in 1827 in six volumes. This was the first atlas of the whole world with sheets at the same scale and on the same projection, making it possible (as was done only once) to put together the lithographed sheets on a globe to make one large globe of nearly eight metres in diameter. This format consequently produced some of the largest scale mapping of previously unreported, or little mapped areas.

All of the map makers listed above included detailed and some very decorative maps of the provinces of the Low Countries. To these must be added the name of the Italian Lodovico Guicciardini, who engraved a charming series of maps of the Provinces published by Plantin in 1582.

France

In the sixteenth century France underwent dramatic and turbulent political upheaval and the relatively stable atmosphere which spawned the science of cartography to the north did not exist. Consequently, apart from a few publications, map production in France is seen as of no great importance until the mid-seventeenth century. From that point onwards, as the French nation grew in wealth and power, her contribution to world geography formed the basis of 'scientific' cartography.

Maps of France and its provinces had appeared in the atlases by Ortelius and Mercator and also in a national atlas in 1594, published by Maurice Bouguereau (the reissue by Le Clerc from 1620 is more commonly found). Maps of the continents by André Thevet and maps from François de Belleforest's edition of Munster's *Cosmographia*, both issued in 1575, can also be found.

The year 1600 saw the birth of Nicolas Sanson who, together with his sons, published in 1654 the *Cartes Générales de la Geographie Anciennes et Nouvelles*. The first edition comprised 100 maps, including the earliest maps of North America to show the five Great Lakes. In addition to this folio atlas, four quarto books of maps of each continent were published, also including interesting maps of North America.

Throughout the seventeenth century, the French were particularly active in North American exploration and their maps are accordingly of great interest. However, their theoretical contributions to cartography are also of interest, albeit for their curiosity value rather than accuracy (see page 48).

After the death of Nicolas Sanson in 1677 his maps were re-engraved on a larger scale and published by Alexis Hubert Jaillot in the *Atlas Nouveau* of 1681. These large-scale versions, superbly presented with large, flamboyant title cartouches and other decorative features, and found fully coloured and heightened with gold, are regarded as amongst the finest map publications of any period. The atlas evolved into the *Atlas François*, subsequently published by Mortier in Amsterdam, and comprising over 150 maps. During this same period Pierre Duval (Nicolas Sanson's nephew), Jean Baptiste Nolin (father and son of the same name), and Nicolas de Fer were active in producing atlases and maps of all parts of the world.

Two other figures, at the end of the seventeenth century, proved of the greatest importance at this stage of scientific cartographic development. These were Jean Dominique Cassini and Guillaume de L'Isle.

Besides the World map for which Cassini is famed (see page 56), his establishment of scientific surveying and triangulation techniques enabled a new and accurate survey of the French coasts to be completed in 1681. After his death his son Jacques commenced in 1733 a survey of the whole of France which, though not published until 1789, was one of the greatest of all cartographic achievements to that date.

The best known figure in French cartography is Guillaume de L'Isle who, having learnt geography from his father, was considered an infant prodigy. He had studied maths and astronomy under Cassini and his first atlas was published in 1700 when he was only 25 years old. His maps of the newly-explored parts of the world reflect the most up-to-date information available and, unlike some of his compatriots, did not include

Theatre de la Guerre en Espagne et en Portugal

One of the most decorative maps of the Iberian peninsula ever produced.
By Pierre Mortier.
Amsterdam c1710
2 sheets, together
120 x 95cm 47 x 37.5in

This superb map in bright original colour was included by Pierre Mortier in his publication of the magnificent Sanson/Jaillot atlas. The unmatched colouring on the two sheets is unfortunate but the eye is drawn to the magnificent titlepiece.

The map is dedicated to King Charles III of Spain in commemoration of the Spanish War of Succession in which the French had been thwarted in their attempt to occupy Spain. Under the command of the King the Spanish army is seen embarking onto a massive fleet lying at anchor. Soldiers and noblemen gaze in admiration at the imposing figure of the King, mounted on a white horse; above the King's head a trumpeting angel and cherubs rejoice at this magnificent scene.

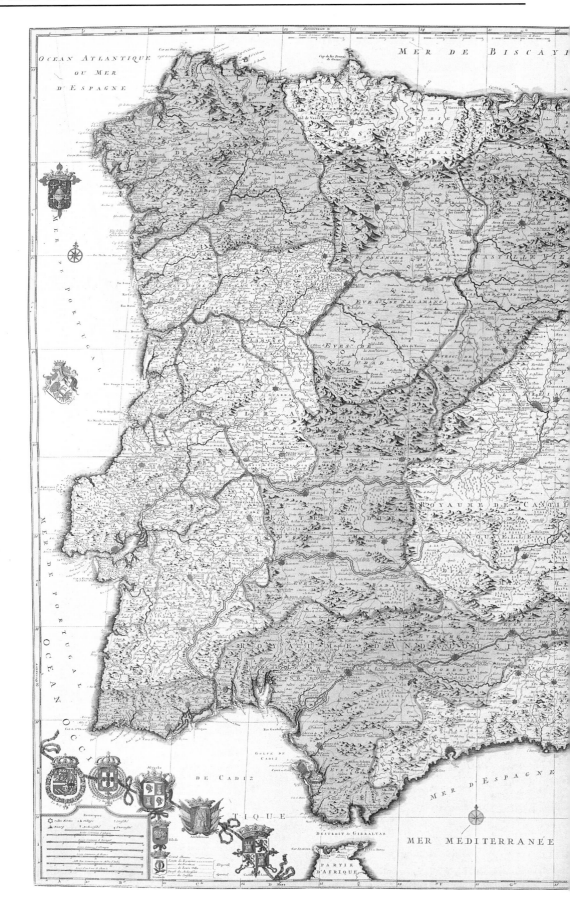

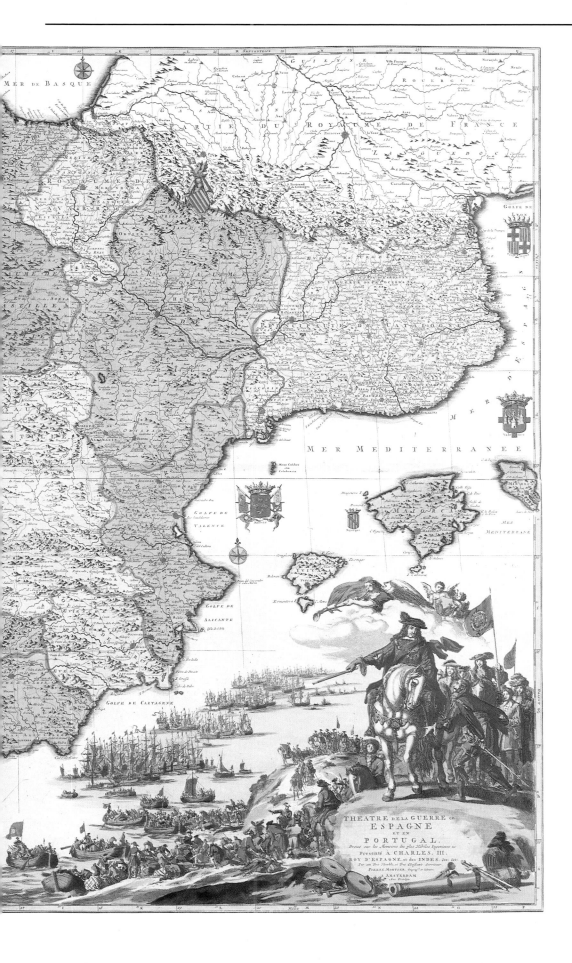

fanciful detail in the absence of solid information. De L'Isle died in 1726 but his maps continued to be republished by Covens and Mortier in Amsterdam, and also by his nephew Philippe Buache in Paris. Buache, unlike his precise uncle, was one of the most prolific practitioners of theoretical cartography and produced a quantity of maps of supposed land formations (particularly with relation to the North West of North America and the great Southern Continent), as well as maps of physical geography (see page 178).

Guillaume de L'Isle was not an only child – his brother, Joseph Nicolas spent 22 years working in, and finally as Director of the Royal Observatory in St Petersburg where he prepared his *Atlas Russicus* of 20 maps, published in 1745.

By mid-century, the tradition established in Paris was being maintained by Jean Baptiste D'Anville whose maps of the continents and whose atlas of China were the best maps available of their period. The continent maps were reissued in numerous French and English and even German and Italian editions.

Throughout the eighteenth and nineteenth centuries numerous atlases were produced or published by French map makers and I list below those whose maps are most likely to be encountered, with the briefest indication as to the style of those maps.

N de Fer Oblong folio maps from the *Atlas Curieux* (c 1700–30). Detailed maps of all parts of the world.

J Chicquet Quarto, general continents and details of Europe (c 1720).

H du Sauzet Quarto, general continents and details of Europe (c 1735).

G Le Rouge Quarto atlas of all parts of the world – including separate West Indian Islands (c 1750). Reissued French editions of some notable English maps of North America.

J N Bellin Quarto atlas maps of ports, harbours and coastlines from *Le Petit Atlas Maritime* (1764). Quarto atlas maps of all parts of the world from Prevost's book of voyages and other publications (c 1750 on). Very large atlas folio maps and charts from Neptune François of all parts of the World (c 1753 on).

G and D Robert de Vaugondy Quarto atlas maps of all parts of the World (1748 on). Large folio maps from *Atlas Universal* (1757, 1783 and 1793). Small folio maps from 1762 until after 1800 when published by Zannoni, De la Marche and others.

R Janvier, Brion de la Tour, C de Mornas, J B Clouet Folio and small folio atlases with maps of all parts of the world (c 1760–70).

R Bonne Quarto atlas maps, rather plain, including maps of newly discovered Pacific Islands (c 1780–90).

V Levasseur Oblong folio maps of the world, the Continents and French Provinces, bordered by elaborate pictorial decoration (c 1840).

Italy and Germany 15th and 16th centuries

Before looking at each country's output separately, it is appropriate to view the influence of the High Renaissance reflected in the output of geographical atlases. Southern Germany and Italy both enjoyed great cultural activity and it is no surprise that, with the publication of various editions of Ptolemy's *Geographia* (see page 34), new maps of different parts of the World should appear.

The development of new printing techniques in Strassburg led to the earliest departures from the Ptolemaic atlas format. Separate sheet maps were being issued by around 1500, and by 1513 Waldseemüller's atlas had styled the *Geographia's* emphasis from the ancient to the modern World. However, Germany's contribution to cartography during the 1500s was, with a few exceptions, limited to the work of Waldseemüller and of Munster, whereas Italy's was larger and more varied.

Italy's location, in the centre of the Mediterranean assured its importance as a maritime, and hence, chart making nation. Traditionally, this prominence was reflected in the production of manuscript charts but with the introduction of copperplate printing, the engravers of Venice and Rome were able to exercise their skills to considerable advantage.

Italy after Ptolemy

There are three periods of Italian cartographic history which are of relevance for the collector: the sixteenth century, the late seventeenth century and the late eighteenth century.

The earliest of these periods saw an active map and chart making industry which was subsequently overshadowed as the Low Countries became Europe's leading map making area.

The importance of Italian chart makers in the history of early cartography led to the production of the first form of 'sea atlas' – the popularly called *Isolario* island book. Several versions of these pilot guides were printed from about 1485 when the first was issued, with rather crudely designed woodcut maps of the Greek Islands, by Bartolome Dalli Sonetti. The most successful, and best known today, was the *Isolario* of Benedetto Bordone, first published in 1528. This included a World map (see page 58), other maps of the New World and newly discovered, or reported, Asian Islands. Giovanni Camocio's *Isole Famose* (c 1574) and Tomaso Porcacchi's *L'Isole piu Famose del Mondo* (1572) were also island books, although their content was of less practical use than the earlier versions – maps of all the major Greek and Mediterranean islands can be found in these, in addition to the British Isles, Iceland, and certain West Indian and South East Asian groups.

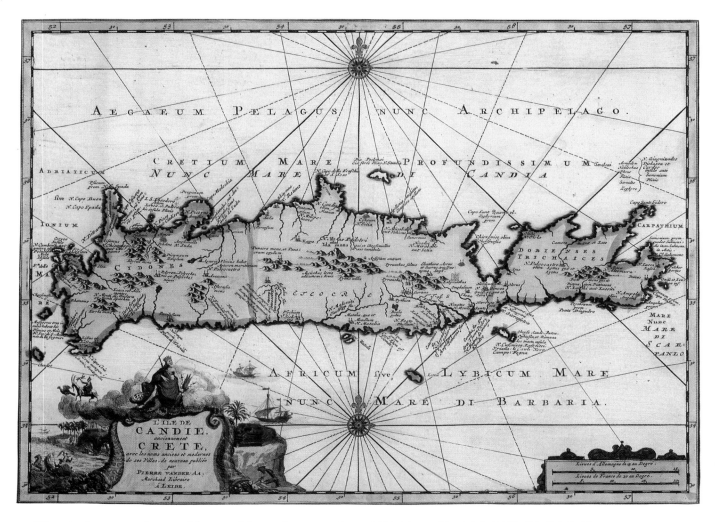

The single most influential figure of the mid-sixteenth century in Italian map production was Giacomo Gastaldi. As Cosmographer to the Republic of Venice, his output included maps of the World (1546), and the Continents of Africa, Asia and Europe (around 1560). He also produced the maps for the 1548 'pocket' atlas of the World, incorporating three new ones of the New World, and the maps in Ramusio's *Voyages* (see page 136). With the exception of certain maps in the *Voyages*, his work was of the most elegant copperplate engraving and typifies much of that being produced during this period.

Fine map engravers were working in Venice and Rome and it was from here that the term 'Lafreri' atlases evolved in the 1570s. Around the 1550s and 1560s the many loose sheet maps engraved by craftsmen such as Gastaldi, Lafreri, Bertelli, Camocio, Duchetti, Forlani, Porro, Tramezini, Zalteri, Zenoi and others, were bound into composite atlases. Of no constant composition, these atlases lacked printed title pages or contents lists. However, on account of a printed title page, produced in Rome and a printed list of maps produced by Lafreri, his name has become generally associated with this sequence of maps.

Despite the quality of Italian cartographic output,

Above
Candie .....

A decorative map of Crete by Pieter Van der Aa.
Leiden 1729
34 x 25cm 13.5 x 10in

This decorative map, in bright original colour leans more towards Crete's classical past than to a modern representation of the island and the location of its villages.

the nation which had produced some of the greatest navigators and explorers (amongst them Marco Polo, Columbus, Vespucci, Cabot and Verrazano), and which had assumed enormous wealth and power throughout the ancient world, was now left behind by the flow of world events, i.e., the accelerating interest in the world beyond Europe and the Middle East, the importance of Europe's Atlantic coastline and the ascendancy of northern European powers.

In 1620, in the wake of other European publications, Italy's first national atlas was published. In contrast to the previous century's plentiful supply of map makers, it is significant that an Englishman, Benjamin Wright, was employed to engrave ten of the maps for this atlas prepared by Giovanni Magini and published three years

Candia

Vincenzo Coronelli's detailed, decorative
map set in a foliate surround.
Venice c 1690
119 x 46cm 47 x 18in

*Engraved on two sheets, this very decorative map appeared first
in Coronelli's 'Atlante Veneto' and is surrounded by an olive
branch framed design on the fruit on which are engraved the
ancient names of the island's cities and towns.*

after his death by his son Fabio. Conversely, the
Englishman, Sir Robert Dudley's sea atlas *Arcano del
Mare* was beautifully engraved by an Italian, Antonio
Lucini, and published in Florence in 1646, and after
Dudley's death in 1661 (see page 60).

The end of the seventeenth century saw a burst of
Italian cartographic activity with Vincenzo Maria
Coronelli at its forefront. Coronelli's initial fame was
through his globe making activities, producing for Louis
XIV a pair of globes, painted rather than printed, of
some 15 feet in diameter. During his career he produced
globes of several different sizes and, subsequently, issued
the gores in book form in 1693 in *Libro dei Globi*.
Coronelli, a member of the Franciscan order, became
a particularly influential figure and, as Cosmographer
to the Venetian Republic, founded the first geographical learned society, *Academie Cosmografica degli Argo-

nauti*, in 1680. His actual cartographic output was
limited to a ten-year period around 1690 but, during
this time, his atlases, charts and town books included
more than 400 different maps of all parts of the world.
His association with the French map makers Tillemont
and Nolin ensured him the most up-to-date information, enabling him to produce characteristic, boldly-
engraved decorative maps, many including each of the
continents on two sheets.

From 1690 to about 1700 two other atlases, similar
in style to Coronelli's, were published by Giacomo
Rossi and Paulo Petrini in Rome. Little cartographic
activity is evident during the first half of the
seventeenth century and the next two atlases of note
can make few claims to originality. Around 1750 an
atlas of maps derived from de L'Isle and Tirion was
published in Rome and around 1780 François Santini
issued a two-volume atlas based on that of Robert de
Vaugondy (see page 106) and d'Anville (see page 106).

By 1785 Antonio Zatta had completed publication
of his four-volume *Atlante Novissimo* comprising 218
maps. Amongst these were provincial maps of Europe;
individual maps of Kent, Surrey, Essex and Middlesex;
decorative maps after Captain Cook of Australia and
New Zealand; and a twelve-sheet sectional version of
Mitchell's important, large-scale map of the United
States of 1755. Zatta's maps are characterized by their
beautiful, decorative titlepieces, often allowing great

Peter Apian, astronomer and mathematician, produced World maps and a *Cosmographia* (see page 51). These were published in various Central European cities.

Very few of these globes and predominantly loose-sheet maps are known to have survived. In the case of Waldseemüller's World map of 1507, only one example of the original 1000 to be printed is now known to us. However, it was copied by contemporary cartographers and these versions very occasionally appear on the market today. Interesting World maps by the following German map makers may be found: Gregor Reisch (1503); Munster and possibly Holbein working together on a map in a book by Grynaeus (1532); J C Honter (1546); in addition to variants of those maps by Waldseemüller, Apian and others.

After Sebastian Munster's death in 1552 the *Cosmographia* continued to be published until 1578, after which time the map plates were re-engraved and used until 1628. However, despite this re-engraving, the work was outdated in content and format, and with the proliferation of good quality cartographic work emanating from the Low Countries, German cartographic production had diminished so that during the seventeenth century there was little notable cartographic publication. Before leaving this period though, reference must be made to the following: Georg Braun's and Frans Hogenberg's *Civitates Orbis Terrarum* (see page 90); Heinrich Bunting's curious maps (see page 186); Théodore de Bry's *Grands Voyages* and *Petits Voyages* (see page 164); and Matthew Quad's interesting World atlas maps (c 1600).

The most prominent map producers during the 1600s were the father and son business, the Matthaus Merians. From 1640 until the next century their 21 volumes of *Topographia* were issued, consisting of town views, plans and maps of all parts of the world. Their panoramic town views, invariably finely engraved, form a fascinating record of seventeenth-century Europe.

Soon after 1700, as the Dutch map business subsided, map and atlas production in Germany became rejuvenated. The most prolific publishers were Johann Baptist (father) and Johann Christoph (son) Homann. Johann Baptist died in 1724 and his son in 1730, but the business continued under the title *Homannische Erben* or Homann's Heirs (Latin – *Homannides Heredibus*) and around 1770 it was publishing a 300-map atlas which included detailed town plans, many with panoramas along the lower border. In very similar style and of the same period are the maps by the Seutter (see page 69) and Lotter families.

Several other map makers were active during this period and maps may be found by C and J Weigel (c 1730), by J G Schreiber (c 1740), and, of more interest, by Heinrich Scherer (c 1710) – (see page 18).

artist's licence when related to the area shown.

From 1792 to 1801 Giovanni Maria Cassini published his three volumes of the *Nuovo Atlante Geografico Universale*. Amongst the boldly engraved maps therein – many with large decorated cartouches – are interesting maps of many individual Pacific Islands including Japan and, after Cook, Australia, New Zealand, Tahiti and the groups of the Society, Friendly, Sandwich and New Hebrides Islands.

Germany and Central Europe after Ptolemy

As has already been noted, Waldseemüller and Munster were the most popular influential mid-European map makers during the sixteenth century and both authors' material was reissued a number of times. Additionally, mention must be made of other German geographers whose work is not so well known today but whose influence was undeniable at that period.

1491 The posthumous publication of the first modern map of Germany by Cardinal Nicolas Cusanus.

1492 Martin Behaim, of Nuremberg, constructed a globe, a summary of knowledge of the world prior to the Age of Discovery.

1493 Hartman Schedel's *Chronicle* was published at Nuremberg (see page 65).

c 1500 Erhard Etzlaub developed his road map of Central Europe.

AFRICA

Although this continent's outline was relatively well known by 1500, a few years after the Cape had been rounded, its internal detail remained a mystery until the last century.

Although the north African littoral was well known to the Romans and the north-east hinterland to the Egyptians, the barrenness of the regions inland prohibited European travel within the continent until relatively recent times. The trade routes throughout Asia, established and enjoyed by Europeans, were never opened up through Africa in the same way. Extremes of arid desert or tropical jungle limited inland travel to a relatively narrow band around the great continent's coast, and although the coastal outline of Africa was quite clearly defined by the late sixteenth century, knowledge of the interior was limited to myth and conjecture. Besides the geographical fantasy, stories of

the Empire of Prester John, of gold and gemstones, of pygmies and statuesque warriors, and of extraordinary human and animal beings proliferated in the absence of fact.

As always, there was an element of truth in many of these tales but it was no surprise that Swift's epithet 'so geographers in Afric maps ...' (see page 26) should be directed at this continent.

Early maps

The first map of the whole continent generally available to collectors is a woodblock by Sebastian Munster (see below), first published in 1540 in the *Geographia ...*

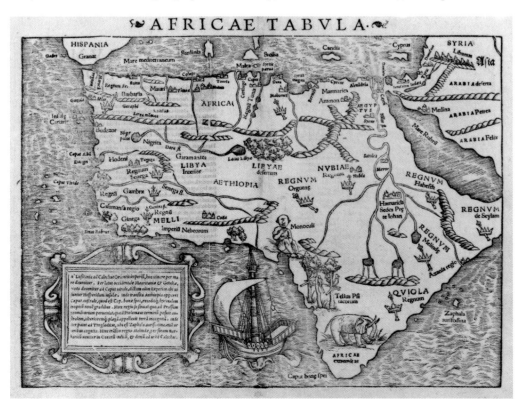

Left
Africae Tabula

Sebastian Munster's important first map of the continent.
Basle 1540
34 x 25.5cm 13.5 x 10in

This well-known woodblock map is probably as famous for its 'monoculi' (one-eyed man) as for the fact that it is the first map of the continent which collectors can reasonably hope to acquire.

110

and, subsequently, in his *Cosmographia...*. This famous map incorporates many of those features which typify subsequent maps of Africa. The north coast is Ptolemaic in outline, the west and south-east coasts are from Portuguese sources, and the Arab ports of Quiloa and Melinda (now Malindi) are identified. A cyclops – Monoculi – is shown, as are parrots and an elephant, and the city of Hamarich, the capital of the legendary Empire of the Christian King Prester John, is set at the confluence of four rivers supplying the Nile. Crowns and sceptres indicate kingdoms. Mountain ranges and forests are also depicted, the largest forest being in the centre of 'Libyae desertum', nowadays the Sahara!

Martin Waldseemüller, in 1513, had published the first printed map of the southern half of Africa and had also, in the same atlas, issued a 'modern' map of north-west Africa. Published within 15 years of da Gama's voyage around Africa these maps, based almost certainly on illegally obtained Portuguese originals, were to become the standard delineations for the next 50 years.

During the sixteenth century fine maps of the whole continent were issued by map makers in Italy, the Low Countries, and, of course, Basle, as in Munster's case. It must also be noted that during this period there were very few sectional maps of parts of Africa. With the exception of one atlas, it was not until after 1650 that further regional maps of the continent appeared.

Typically, an atlas might contain a map of the continent, one north-west map (or Morocco and Guinea as separates), the Barbary coast, Egypt, the 'Empire of Prester John' (Abyssinorum, or central

---

Below
Tabula Moderna Primae Partis Aphricae

Martin Waldseemüller's fine woodblock map – the first printed map of this coastline.
Strassburg 1513-20
57 x 41cm 22.5 x 16in

Prior to 1513 any printed maps showing the North African coastline had been based on the ancient Ptolemaic geography. Here, Waldseemüller has made use of Portuguese reports to give more detail and a better outline than before for the western coasts. During the previous century the Portuguese, under Prince Henry, had gradually travelled further south – Cape Bojador, just south of the Tropic of Cancer, had been passed in 1434; Cape Verde was reached in 1444; and, after a halt in voyages, by 1481 the Equator had been traversed; only seven years later, Dias rounded the Cape.

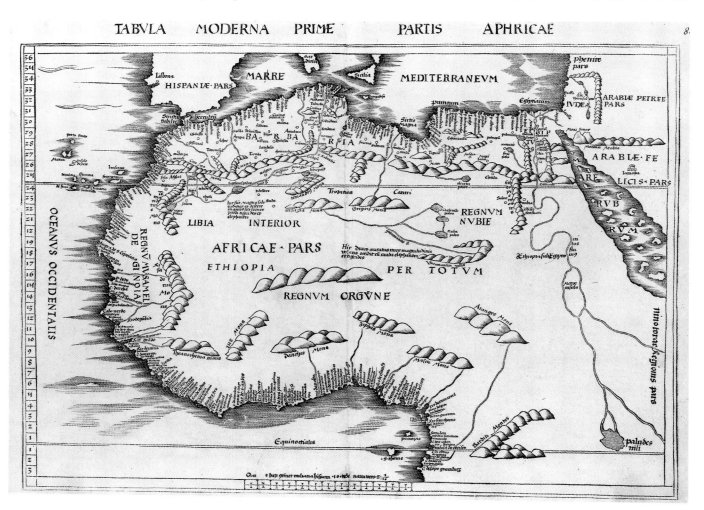

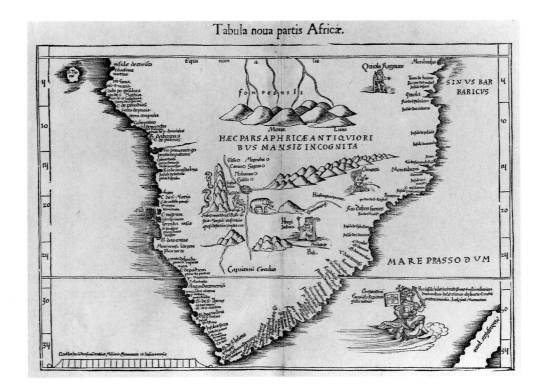

Tabula noua partis Africæ.

HÆC PARS APHRICÆ ANTIQUIORI BVS MANSIT INCOGNITA

Left
Tabula Nova Partis Africae

A fascinating reissue of Waldseemüller's map – the first of South Africa.
Strassburg 1522-41
42 x 30cm 16.5 x 12in

In 1513 and 1520 Martin Waldseemüller had issued his edition of 'Geographia' with, amongst others, a completely new map – the first – of Southern Africa. In 1522 the maps were reissued on a reduced format and with certain additions. This version of the map appeared in 1522, 1525, 1535 and 1541. Under the control of different editors and publishers, editions of single maps may be identified by reference to the authorities.

Left
Cairus, Ovae Olim Babylon ....

Georg Braun and Frans Hogenberg's fascinating plan of the city of Cairo.
Cologne 1572
48 x 33cm 19 x 13in

Based on an Italian engraving of around 1550 this example of the bird's-eye plan is in fine original colour (with gold highlights on garment trimmings). In the foreground are characters in Arab dress, and, on the left, horsemen are racing. On the right can be seen the pyramids and the sphinx. An annotated key identifies further numbered sites around the city.

Africa) and, also possibly, Southern Africa.

With the exception of particularly rare Italian maps of Africa and its ports, the influence of Ptolemy and Waldseemüller dominated the country's cartography until around 1570, in which year, Ortelius's atlas was published and, about which time, the influence of Mercator's World map was making an impact. Ortelius's atlas contained modern, as opposed to historical, maps of Africa, Morocco, and, from 1573, the Empire of Prester John. The Mercator/Hondius atlas of 1606 added to these three a map of Guinea, Barbary and also

Jodocus Hondius's own map of Africa, complementing Mercator's original of 1595.

In 1635, Willem Blaeu produced versions of maps of Morocco and 'Prester John' and a new, elegant map of Southern Africa which was copied by Jansson and had a publishing life through the next eighty-odd years. In addition Joan Blaeu, in volume X of the *Atlas Major* (1662), incorporated new maps of Barbary, Egypt, Madagascar, the islands of Malta and Gozo, the Cape Verdes, the Canaries, 'Nigeria', and the Congo and Angola.

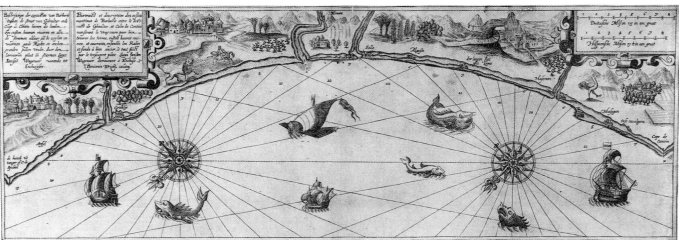

Beyond this very standardized pattern of regional mapping there are two other publications of particular note at the end of the sixteenth century. The first atlas of the continent was published in Venice in 1588. Livio Santo was a mathematician, cosmographer and instrument maker who intended to produce a series of books describing, and depicting in maps, all areas of the world. Unfortunately, he died as the first atlas was being published, and although the dominant cartographic concepts on the 12 maps are Ptolemaic, the fine design and elegant engraving by his brother Guilio make these uncommon maps superb examples of Italian map making.

The other major cartographic production, not fully appreciated at the time, is an interesting map of the greater part of the continent by Fillipo Pigafetta. This finely engraved map revised the Ptolemaic tradition of the Nile rising from two lakes side by side, and offered a truer picture of the relationship between Lake Victoria – the true source – and Lake Tanganyika. Pigafetta's map accompanied the reports of a Portuguese explorer, Duarte Lopes and also gives the most up-to-date and correct depiction of the Congo and central African regions. In fact, another map by Pigafetta shows this area in detail. Pigafetta's maps

Above
The earliest detailed charts of the N W Moroccan coast

From Lucas Jansz Waghenaer's *Thresoor der Zeevaert*.
Leiden c1606
each approx 53 x 20cm 21 x 8in

These beautifully engraved charts are amongst the earliest non-European charts to have been printed. They were added to Waghenaer's excellent chart book to aid Dutch seamen on their course for the Canaries and, subsequently, to southern seas. The engravers' signatures – Joshua van den Ende and Benjamin Wright – can be seen and the differences in their styles compared. Both lean heavily on the styles of earlier manuscript charts, or portolans, though the Englishman Wright's work is more illustrative and the Dutchman's more formalized.

appear in editions of Theodore de Bry's *Petits Voyages* (1598), and in a re-engraved version in the English text edition of Hugo van Linschoten's *Itinerario* of 1598. Unfortunately, few map makers accepted Pigafetta's realignment of the Central Africa Lakes and his map was not as influential as others.

In Linschoten's book of voyages (see page 120), there

113

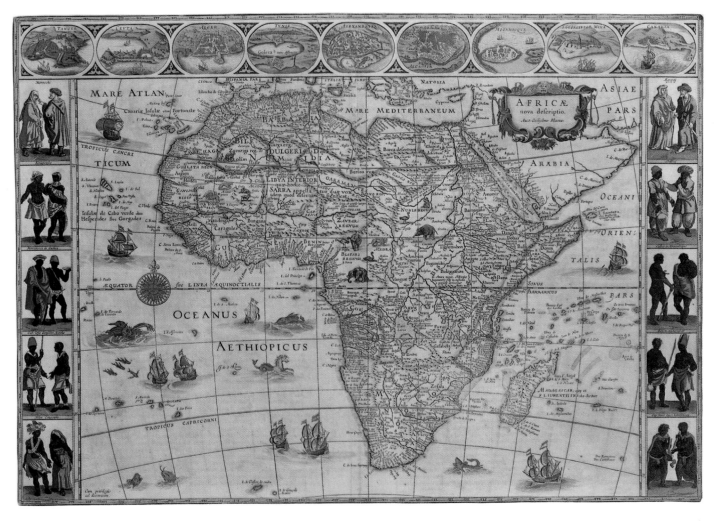

Above
Africae

Guillaume Blaeu's famous 'cartes à
figure'
Amsterdam 1619-c1650
56 x 41cm 20 x 16in

*This famous decorative map is bordered on each side by panels
of native characters and, above, oval plans and panoramas of
Tangier, Ceuta, Alger, Tunis, Alexandria, Cairo, Mozambique.
George El Mina, and the Canary Islands. Such a selection of
views represents an interesting selection of 'Old World'
Mediterranean and Egyptian Cities combined with the early
Portuguese and, subsequently, Dutch interests on the Atlantic
and African coastlines.*

*This map was originally engraved, by Blaeu about 1618, his
signature 'Guil Janssonis' in the titlepiece identifying the issue.
However, in 1630 the family nick-name 'Blaeu' replaced
'Janssonis' to avoid confusion with the rival firm of mapsellers
active in Amsterdam. The early state of this map is very rare since
it predates the first issue of Blaeu's land-atlas and copies of the
map would have been issued separately or in composite atlases.*

*The early colour on this map is slightly heavier than is usually
the case with Blaeu. Possibly the map was not coloured by the
Blaeus 'in-house' but the owner commissioned his own colourist
to do the work. This colour is more typical of that of Hondius or
Jansson.*

are several other maps which merit mention. A pair of
exceptionally decorative map engravings show, respectively, the coastline south of Guinea to the Bay of
Algoa, and the coastline north and east from Algoa to
beyond the Equator. Each map is decorated in the most
flamboyant, though elegant style with large rococo
strapwork and cartouches, ships, monsters, ornate
compass roses, and so on. In the same publication, and
of relevance to Africa, are some particularly fine maps
and views of some of the Atlantic Islands – Saint
Helena, Ascension, Angora, the Cape Verdes and
Tercera. These islands served an essential function as
places to shelter, revictual and obtain fresh water and
were of the greatest importance to South Atlantic
shipping.

As we have seen, during the first half of the
seventeenth century relatively little advance in the
continent's mapping was achieved. However, from the
second half, maps by Blaeu and by Nicolas Sanson,
published in Amsterdam and Paris, and by Dapper and
Ogilby, can be found. Cartographically, there is little
new in these maps but their contrasting styles and
variations in decorative content make them of interest
to the collector. In London, John Speed's atlas
contained a particularly decorative, bordered map of

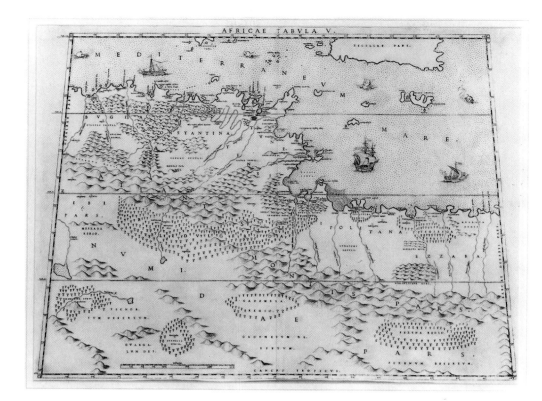

AFRICAE TABVLA V.

Africae Tabula V

Livio Sanuto's elegantly engraved map of part of the North African coast.
Venice 1588
50 x 39cm 20 x 15.5in

Sanuto's rare atlas of the continent is important as the first devoted solely to Africa, but his reliance, in the main on Ptolemaic cartographic theory diminished its impact.

However, one can see the fine quality and design of Italian copperplate engraving of the period, the maps being engraved by Sanuto's brother Giulio.

Livio Sanuto's intention had been to produce atlases of each of the continents – unfortunately his ambitions were not to be fulfilled since he died before the first volume of his atlas was published.

Africa and throughout the century decorative maps were produced.

At the turn of the seventeenth century, despite the notable chart production of the Mortiers and Van Keulens, the major map producing areas diverge to England and France. In Paris the map making houses of de L'Isle and, subsequently, d'Anville produced maps sacrificing decoration and fancy to detail or no detail depending upon what was known. In London John Senex, Herman Moll, Emanuel Bowen and others produced good quality world atlases and instituted an English map making tradition that extends from around 1700 throughout the period of our interest. Bowen, in fact, based his maps on those of his French contemporaries. Thomas Kitchin, Robert Laurie and James Wyld all produced good, scientifically-based maps recording a period when English map making techniques were maturing and British colonial activity was most active.

ASIA

The size and diversity of the Asian continent is matched by the number and range of different map groupings available to collectors within this geographical area. Accordingly, in this chapter, a brief discussion of some of the more general maps is followed by broad cartographic histories of those Asian areas most popular with collectors today.

Within the Asian continent the map making tradition predates European printing by hundreds, even thousands, of years. Parts of the Middle East, India and the Orient can claim early cartographic artefacts but these invariably unique items are well beyond the scope of this book.

Ptolemy's *Geographia* included some 12 maps of parts of Southern Asia, of which the majority depict the Near and Middle East. The Roman road map, *Tabula Peutingarum*, encompassed the same area as *Geogra-*

phia, but extended further to South East Asia and the island of Taprobana (sometimes taken for Ceylon or Sumatra).

The heartland and bulk of Asia, like Africa, consists of most inhospitable terrain and so the areas of greatest cartographic interest lie south of the Caspian and the Himalayan barrier and inland from the south and eastern coastlines.

In 1513 Martin Waldseemüller produced, in his edition of *Geographia*, a new map of southern Asia.

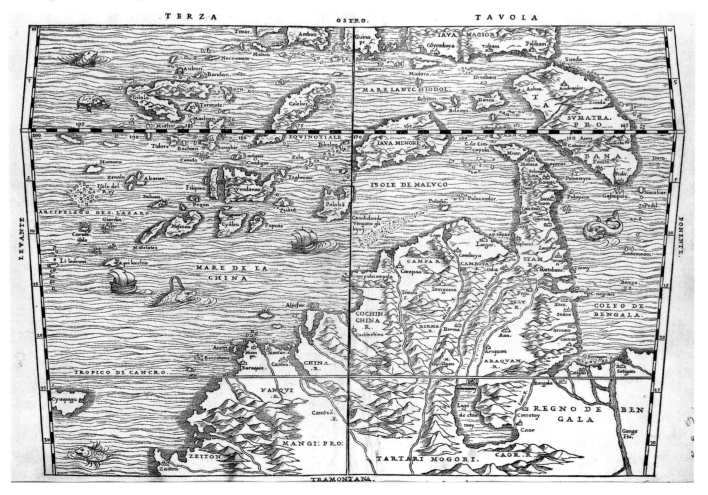

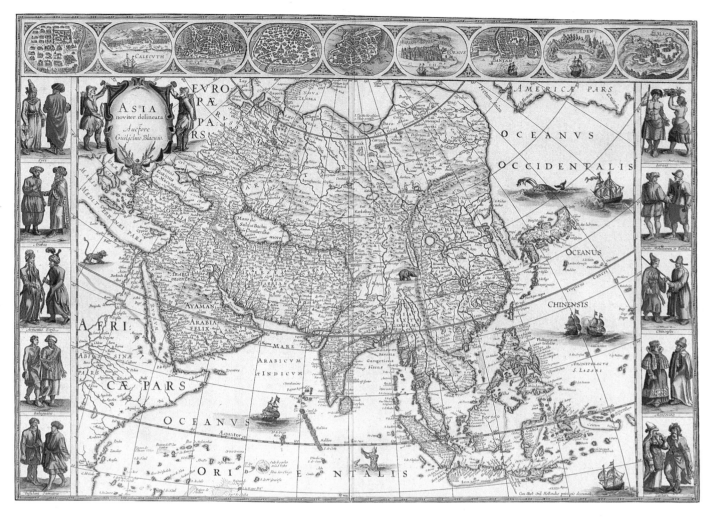

Above
Asia

Guillaume Blaeu's famous map of the
Continent.
Amsterdam 1618-c1650
55.5 x 41cm 22 x 16in

*A fine engraving in delicate original colour, typical of Blaeu.
Panelled borders at each side show costumed figures from various
parts of Asia – Syria, Arabia, Persia, China, Muscovy, Tartary
and South East Asian Islands. Views along the top border are of
Canton, Calicut, Goa, Damascus, Jerusalem, Ormus, Bantam,
Aden and Macao – all important trading places.*

Left
Terza Tavola

A curious map from Ramusio's *Voyages.*
Venice 1554
39.5 x 27.5cm 13.5 x 11in

*An unusual map from Ramusio's important three-volume book
of voyages published in Venice. Three maps by Gastaldi in
Volume I cover Africa, the Middle East and South East Asia and
are printed with south at the top of the page.*

*This was one of the most detailed maps of South East Asia at
the time and is one of the earliest to approach the correct location
for Japan – here named 'Cympagu'.*

Taking in the Persian Gulf, India and South East Asia,
this map identifies only coastal locations and would
have been based – like the two corresponding maps of
Western and Southern Africa – on Portuguese accounts. Waldseemüller's atlas appears in a second
edition (1520), and two years later in a reduced format,
also published in Strassburg. This issue of *Geographia*
appeared in four editions (1522, 1525, 1535 – Lyons
and 1541 – Vienne), and added two further maps of
parts of Asia (one of the South East and one of China
and Japan). These crude maps are the first to
concentrate on these particular areas but are not as
accurate as they might be and are influenced extensively
by details from Martin Behaim's globe of 1492 (the
oldest surviving globe) and Marco Polo's reports.

The first map of the continent of Asia appeared in
Munster's *Geographia* in 1540, and although the map
extends to the eastern seaboard of 'Cathay' and
identifies 'Archipelagus 7448 Insularis', Japan is not
included. Again this map was based on Marco Polo's
reports which cited an archipelago of 7448 islands (one
wonders who had counted them), and Japan was still
estimated as too far east, in fact appearing on Munster's
map of America.

With the publication of Ortelius's atlas in 1570 and

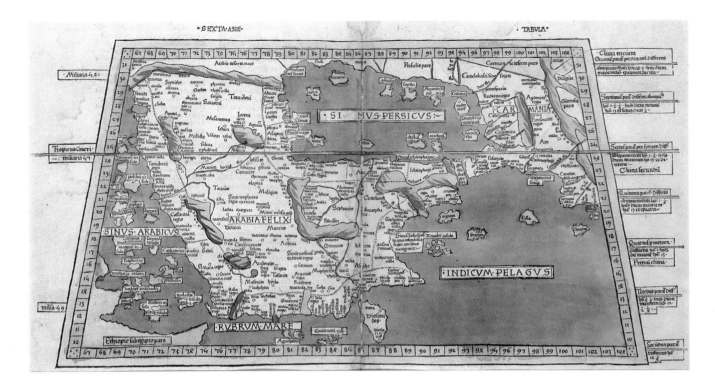

Above
Sexta Asie Tabula

The first woodblock version of Claudius
Ptolemy's *Arabia Felix*.
Ulm 1482-86
55 x 30cm 21.5 x 12in

*This boldly engraved woodblock map shows the ancient Ptolemaic
shape for today's Saudi Arabia. Islands in the seas are exaggerated
in size and the features most prominent on the land are
mountainous regions.*

*The title refers to this map as the 'sixth table (or map) of
Asia', this being one of twelve maps of ancient Asia from the
'Geographia'. This atlas was published twice and maps from it
are amongst the earliest which might be found in contemporary
colour (see page 34); typically the edition of 1482 will have a
strong blue colouration on the sea, whereas the second has an
'olivey' greenish-brown wash on the sea. Furthermore, the maps
of the 1486 edition have the addition of titles added to the top,
as here, and also, the text on the reverse of the sheet was reset
without the decorative woodcut borders of the first edition.*

contributed maps to Giovanni Ramusio's *Raccolta di
Navigationi et Viaggi*, published in Venice from 1550.

Ramusio's *Voyages*, as it was known, is regarded as
the first of the great travel compilations. In three
volumes published between 1550 and 1556, Ramusio
brought together the most accurate and up-to-date
reports of many of the most recent and major older
voyages of discovery and exploration. The most
extensive early reports of Marco Polo in Asia, of the
Portuguese Lopes and Barbosa on the Indian Ocean,
of Pigafetta in Africa (see page 113), and Verrazano in
North America (see page 141) are here. The book also
included maps of each of these areas by Gastaldi (see
page 116). Of equal interest for its maps, and for its
great contemporary importance, is the work of Hugo
van Linschoten whose *Itinerario* (Amsterdam 1596,
with an English edition two years later), was produced
to inform and stimulate interest in the newly established trading colonies. Covering the Indies, Indian
Ocean coast of Africa, the Atlantic coast of Africa and
South America, the book is a compendium of sailing
directions, information on islands, harbours, local
produce and so on – in general terms, a compendium
for any merchant wishing to trade in these exciting and
active new markets. The book includes fine maps
engraved by the van Langren and Doetechum families
and, in the English editions, by Robert Becket.

These maps are amongst the most spectacular of any
period and mark an exciting point in the development
of map making. Prior to this time maps were used
almost exclusively as separate sheet publications, or
within the context of an atlas. A world map might

Mercator's in 1595, the outline and internal detail of
maps of the continent improved considerably. Moreover, in these atlases, maps were added showing
individual countries and areas, thus allowing for much
greater detail than before. Around 1560 Giacomo
Gastaldi had produced a finely engraved three- or
four-sheet map of Asia – the best of its time – and this
provided much of the information which was subsequently used by Ortelius and Mercator. Gastaldi, a
prolific map maker, included separate maps of parts of
Asia in his important small atlas of 1548 and also

appear in conjunction with a world history but the great majority of printed maps were bound into an atlas. In the travel books of Ramusio, Linschoten and de Bry (see pages 136, 120, 164), we see maps used to illustrate detail within the text and, particularly in the case of the Dutch maps for Linschoten, to inspire and excite active interest in the persuance of trade in these new markets.

Throughout the seventeenth and early eighteenth centuries the continent inspired map makers to produce decorative and informative maps, often with figured borders (see page 26) or with large title cartouches, incorporating costumed figures of the mystic and exotic countries of the Orient.

Arabia

Excluding the Ptolemaic maps of the Arabian peninsula and Waldseemüller's 1513 map of southern Asia, the first 'modern' map of Arabia appears in Gastaldi's atlas of 1548, which was reissued in a slightly larger format in 1561. Two other maps by Gastaldi – one a separately-issued, detailed and very fine map of Arabia, (taken from his larger map of Asia, c 1560), the other, in Ramusio's *Voyages* – and a miniature map from

Barent Langenes' *Caert–Thresoor* (1598) are the only sixteenth-century maps of just the Arabian peninsula.

From 1570, the *Turcicum Imperium* (the Turkish Empire) map from Ortelius's atlas proved to be the more common format, incorporating the Ottoman

---

Below
The Turkish Empire

John Speed's decorative map.
London 1627-76
51 x 39cm 20 x 15.5in

The whole Middle East is shown on this decorative map. Figured borders have full-length portraits of Greek, Egyptian, Assyrian, and Persian men with their wives. Town plans at the top include Famagusta, Damascus, Alexandria, Jerusalem, Constantinople, Cairo and Ormus.

Annotations at any blank point on the map provide historical and geographical details of countries shown. Speed's maps are, of course, well known for their descriptive text, in English, on the reverse further describing the countries shown.

N.B. The date '1626' on the titlepiece is that of the engraving and not of publication. This map was first published in 1627 but this edition is identified by the names of the publishers – Thomas Basset and Richard Chiswell – as 1676.

Linschoten's *Itinerario*

Jan Huyghen van Linschoten's *Itinerario* is seen as one of the greatest travel books of any period. The book documents much information concerning the Portuguese colonies and their trade at the point when both the Dutch and English maritime trading nations were becoming established and capable of competing for trade, which had, hitherto, been controlled strictly by the Portuguese. Portugal's resources – particularly of capable manpower – were extended beyond that physically small country's limits.

Linschoten had spent several years in India where he obtained much of his information for both the text and cartographic content of his book – the book was of immense benefit to the merchant marine for its detail on trade and navigation throughout the Indies and ran to several editions: Dutch in 1596, 1605, 1614, 1623 and 1644; English in 1598; Latin in 1599; and French in 1610, 1619 and 1638. The maps accompanying the volume are of particular interest since they provide detail based, in the main, on Portuguese charts which were, of course, never otherwise printed.

The World map usually associated with this publication is the magnificent double hemisphere of Petrus Plancius. Besides maps, views and coastal profiles of the major Atlantic Islands – the Cape Verdes, Ascension and Saint Helena – and a plan of the Portuguese base in India, Goa, there are fine, detailed maps of the following areas: west and southern Africa; southern and eastern Africa; the Horn of Africa, Arabia, Persia, and India; south-east and oriental Asia from Malaysia to Japan. (Some editions included a map of the Moluccas – the Spice Islands.) It will be apparent that the four maps above provide a complete coverage of the coastlines from western Africa to the Orient – essential for any East India bound vessel.

For the English edition, published by John Wolfe, new plates were cut by Robert Beckit and in the 1599 edition a spectacular map of William Barent's voyage to the Arctic is included.

A striking map detailing the Portuguese and Spanish Empire in South America is also found in all editions.

In addition to the importance of their cartographic content these fine maps are amongst the most lavishly designed and engraved of any period and represent Dutch mapmaking at its most stimulating period.

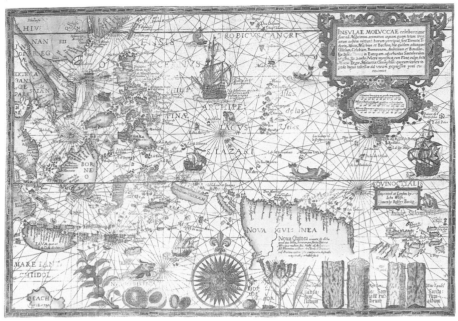

Exacta & Accurata ... orarum maritimarum ...

South-east and oriental Asia from Linschoten's volume.
Amsterdam 1596
53 x 38cm 21 x 15in

A famous and important map based on the work of Portuguese cartographer Fernao Vaz Dourado. This was the best available at the time although the 'shrimp-like' outline of Japan is particularly curious. The orientation of north to the left is also unusual and allows for the inclusion of a landmass called 'Beach' at the right, or southern side of the chart. The name 'Beach' derives from Marco Polo who applied the name to a great southern continent about which he was told when visiting China in the thirteenth century.

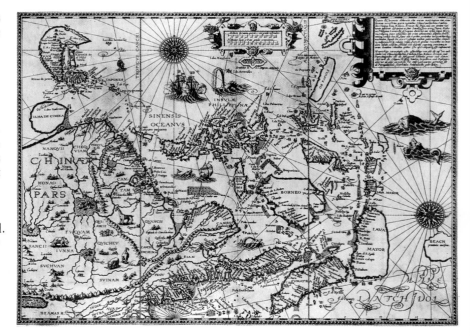

Insulae Moluccae ...

A spectacular chart from the English edition of Linschoten's work.
London 1598
53 x 38cm 21 x 15in

After a Dutch original by Plancius and engraved by Johannes a Doetechum, this fine map shows the Spice Islands at a detail never previously used. The illustrations along the lower margin are of local products – nutmeg, mace, cloves and sandalwood.

Empire from Constantinople to Aden and Ormus.

Decorative, detailed maps of the Turkish Empire can be found produced throughout the seventeenth and eighteenth centuries, though, from about 1660 onwards there were maps produced by Sanson, Blaeu, Jansson, Blome and others concentrating on Arabia Felix – the peninsular area known generally today as Saudi Arabia.

As might be expected, internal detail was not very accurate as few Europeans ventured away from the trading ports until the nineteenth century. However, the maps of Sanson (1652) and subsequently, de L'Isle (1700), and d'Anville (1751) can be seen as distinct improvements on their predecessors.

The Holy Land

Given the historical significance and classical interest of this pivotal part of the Ancient World, it is hardly surprising to note that there were probably more maps made of this region than of any other in the world.

Besides medieval World maps which showed Jerusalem at the centre, maps of the Bible lands appeared not just in atlases but also in Bibles, histories of the world and its peoples, or as separately-issued wall maps

designed solely to instruct or for decoration. Furthermore, many atlases from Ortelius onwards included more than one map of the region showing, under a variety of titles, different historical periods or events.

The *Theatrum* of 1606 included five maps of immediate relevance: two by Tilleman Stella –

Below
Palestina Sive Terrae Sanctae Descriptio

Jan Jansson's decorative, sought-after map of the Holy Land.
Amsterdam 1630
55 x 44cm 22.5 x 17.5in

The map content of this decorative engraving is based on the standard Adrichom cartography of 1592. The most striking feature, however, is the series of eighteen scenes of Old Testament history which are shown at top and bottom and the inset view of Jerusalem to the left.

This plate appeared in the 'Atlantis Maioris Appendix' of 1630 and 1631 and later in a few editions of the historical atlas by Georg Horn which was issued by Jansson. However, it was not often used and consequently is difficult to find on the market today.

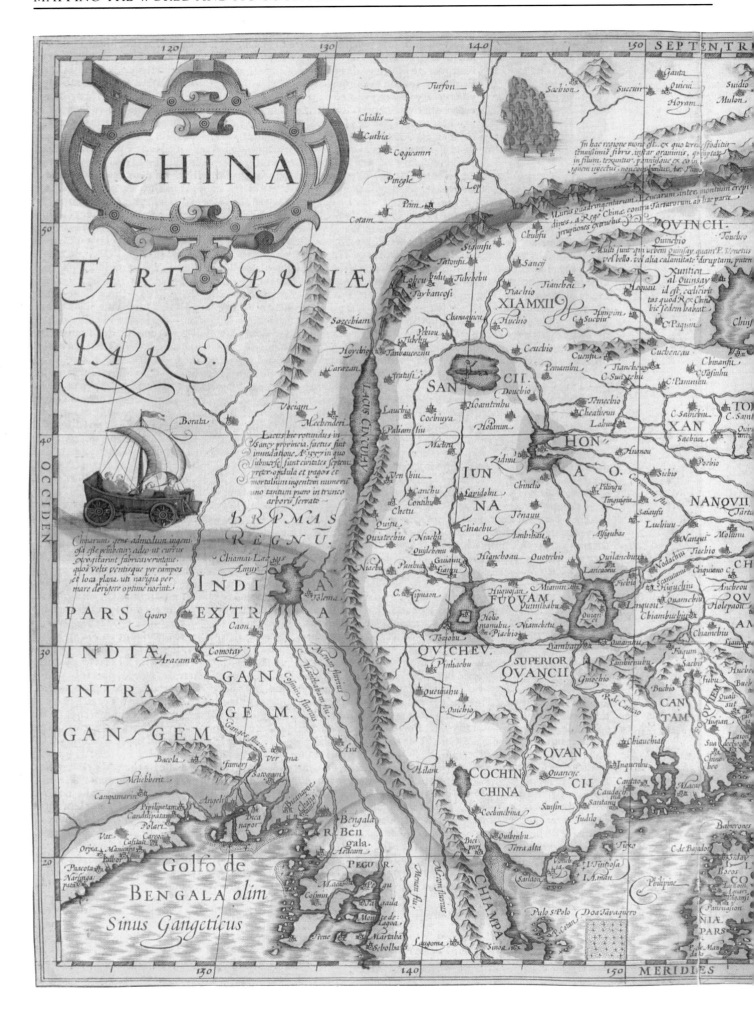

China

A decorative, fascinating map of China and Japan by Jodocus Hondius.
Amsterdam 1606
46 x 34cm 18 x 13.5in

A boldly engraved map in good early colour showing Korea as an island, and combining elements of Barbuda's map of China and Teixeira's map of Japan (both published earlier by Ortelius). Amongst the decorative features a sea monster, deer on the North American coast, Dutch and Japanese ships, and a Chinese wind-blown land vehicle can be seen. A scene in a panel on the right shows an ancient Japanese torturing technique.

The Mercator/Hondius/ Jansson series of atlases was published in various European language versions. However, one of the most interesting editions is the sole English text edition of 1636.

Although the face of the maps remains constant the text on the reverse is, of course, translated and provides a fascinating insight into early European information of foreign and exotic countries.

The text for this map describes the climate, national features, inhabitants and economy of China:

'In every Mechanick art, and manifacture they are so exceeding skillfull and cunning, that their wockemanship seemeth rather to be made by nature, then by the hand of man... The Portugalls report many straynge things of the subtilitie of their spirits, among the rest, which is most to be wondered at, they have wind-chariots or wagens with sailes and knowe how to steere and governe them with such dexteritie, that in a short time they can carrie themselves where they please...' (see left hand side of the map).

Palestinae Sive Totius Terrae Promissionis and *Typus Chorographicus, Celebrium Locorum In Regno Judea et Israhel*; one by Schrot *Terra Sanctra*; and two by Ortelius himself – *Abraham Patriarchae Peregrinatio et Vita* and *Perergrinations Divi Pauli*. Ortelius's own interest in classical history led him to produce a particularly fine map of Canaan at the time of the Patriarchs, surrounded by some 22 finely-engraved circular scenes of events during that period. This is one of the most sought after maps of the area and is generally known as *Abraham's Travels*.

Christian Van Adrichom's *Theatrum Terrae Sanctae* published in 1592, included a fine large-scale map of the Holy Land and nine other maps showing the location of the tribes of Israel and a detailed plan of Jerusalem. This popular format was repeated by Thomas Fuller in his *Pisgah-Sight of Palestine* published in London in 1650.

Few atlases of this time were issued without a detailed map of the Holy Land. In England a charming miniature map by historian, genealogist and map maker, John Speed, appeared in John Bill's issue of The Bible *c* 1620, his most famous map of *Canaan* having appeared, first, in 1611. This was subsequently reissued in the 1676 edition of his *Prospect* and is the most commonly found form of this map. Innumerable other maps, from the first printed Holy Land map in the Ulm edition of Ptolemy (1482) to the eighteenth century are worthy of detailed attention, but the scope of this book cannot, unfortunately, allow for this. Suffice to say that there are fine maps by all of the famous map makers, and plans and panoramas of Jerusalem by most of the better, or more important, publishers and engravers throughout the period of printed map making.

Persia

As we have seen with Arabian maps, the first 'modern' map appeared in 1548. However, Persia appeared as a sole entity on Ptolemaic maps and has continued to do so throughout the history of map making. Occasionally, the country has appeared with the rest of the Middle East, but it would be quite possible for the collector of Persian maps to build a collection of maps solely of that country across the centuries.

For example, a collection of the following maps would form a good representation of the work of major cartographers and avoid unnecessary duplication in style, visual appearance and date, *viz.* Ptolemy (*c* 1500), Gastaldi (1548), Munster (*c* 1550), Ortelius (1570), Hondius (1608), Speed (1627), Blaeu (*c* 1650), Blome (*c* 1680), de L'Isle (*c* 1700), Seutter (*c* 1740), Bowen (*c* 1750), Santini/d'Anville (*c* 1780), Cary (*c* 1810), Arrowsmith (*c* 1840), etc.

India

Maps of the sub-continent of India, albeit in virtually unrecognizable form, appear in the earliest editions of Ptolemy's *Geographia*. After Waldseemüller's 1513 map and its subsequent issues, Gastaldi's miniature atlas of 1548 was the first to include a map showing India separately. This was revised in a larger format in 1561 and shows a particularly narrow peninsula, whereas his map in Ramusio's *Voyages* (see page 116) shows a notably broader shape.

Ortelius in 1584, and Jodocus Hondius in 1608, included India on their maps with South East Asia while the Linschoten map of 1596 includes it with Arabia and Persia. Detail at this time extended the length of the peninsula's West coast but petered out towards the east and the Bay of Bengal. Information was gleaned primarily from Portuguese sources and inland detail was negligible.

In 1600 the English East India Company was founded and although the Portuguese, Dutch and subsequently French, had extensive trading interests in the sub-continent, it was the English who opened up and ultimately benefited most, as a European Imperial power, from the riches of India.

The most influential map of India of the early seventeenth century was produced by William Baffin and published in 1619, some 11 years after the East India Company's first official expedition led by William Hawkins in 1608 (see opposite).

During the next one hundred years maps of India were generally issued in two sheets – a north sheet according to Baffin's map, and a southern sheet including Ceylon. The first important non-Ptolemaic map of *Taprobana* (the ancient name for Ceylon) was that of 1606 in the Mercator/Hondius atlas. This map was based on an original by the Portuguese, Cypriano Sanchez. The Dutch, in occupation during the following century, produced constantly improved maps of the island, some with large and spectacular title cartouches. A number of superb maps by Van Keulen and Visscher, or the German, Seutter, can be found.

As English and French knowledge of the interior of the sub-continent increased, so the quality of mapping improved. Moreover, the military interests of both European powers necessitated improved surveys and so we find new maps by de L'Isle (1723) and d'Anville (1752), published in Paris and, later in the century, fine, large-scale maps by British map makers. After about 1763 French ambitions in India had been eclipsed, and to ensure military superiority and assess the extent of the East India Company's lands, it was deemed necessary to appoint a Surveyor General. Captain James Rennell was appointed and in 1773 his systematic surveying was completed. During the next several years Rennell compiled and drew maps from the

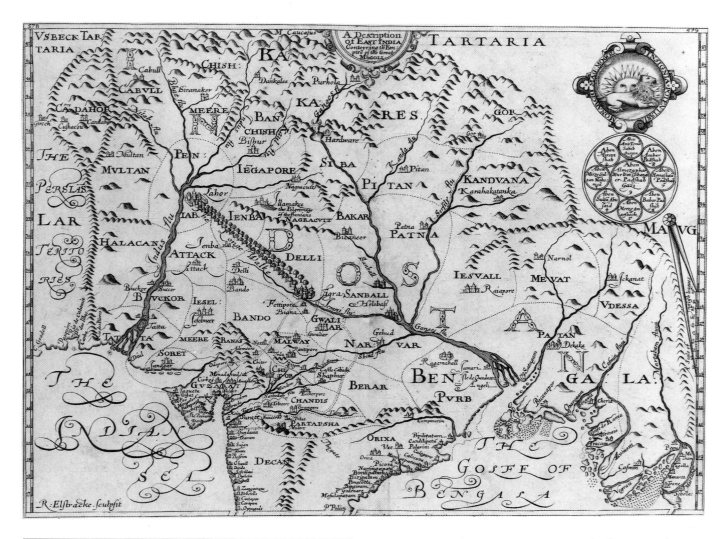

Above
A Description of East India

William Baffin's influential map of the
Mogul Empire.
London 1625
37 x 28cm 14.5 x 11in

An influential and important map of northern India. This map was finely engraved by Reynold Elstracke for Samuel Purchas' great five-volume book of voyages and travels and was based on an earlier, separately-issued map, also engraved by Elstracke, by the much travelled William Baffin.

In 1615 the East India Company, after their earlier attempt at establishing trading connections with India had failed, requested King James I to appoint an ambassador to approach the Emperor Jehangir. Sir Thomas Roe duly spent four years in Surat and, subsequently, with William Baffin — better known for his surveys of the Canadian coast — produced the map which became the standard presentation of the area for the next one hundred years.

rough field drafts he had gathered and in 1779 the first edition of the *Bengal Atlas* with nine maps was published; this was expanded to 23 maps within two years.

The work by James Rennell and the maps of the coasts and harbours by Alexander Dalrymple were

outstanding achievements in a period when British map making was at its most innovative, and their work subsequently formed the basis for most nineteenth-century maps of the sub-continent.

South East Asia

The earliest maps of the area tend towards a conglomeration of various shaped islands, which is of course the case; it was not until the end of the eighteenth century that the coastlines of New Guinea and other islands were defined reasonably accurately. The most common sixteenth-century map is that from Ortelius's atlas, first included in 1584, while the most spectacular maps are those from Linschoten's *Voyages* of South East Asia including China and Japan and of the Molucca Islands (see page 120).

There are few maps of individual islands apart from the Philippine group. Maps of the Philippines appear in atlases throughout the development of cartography, from the miniature atlas of Langenes in 1598 onwards. Maps of individual islands – Java, Bali, Sumatra and Borneo – are sometimes found, as are charts of the Straits of Singapore although these, obviously, do not provide great detail of Singapore as it is known today.

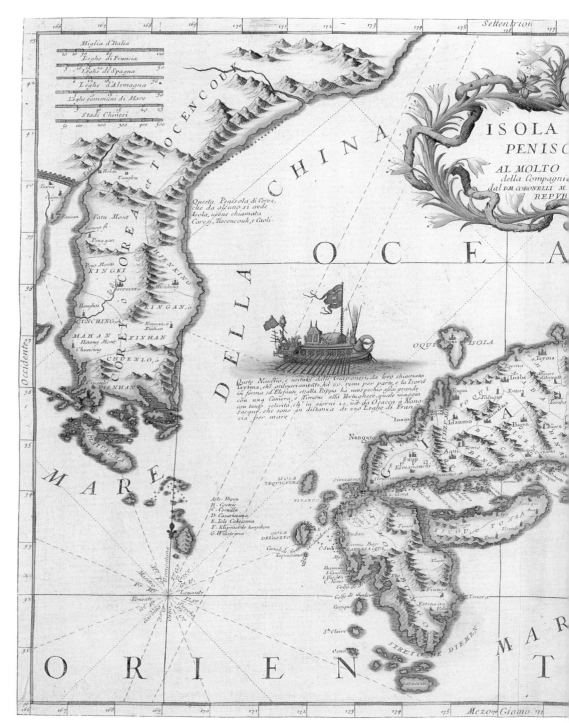

Isola Del Giapone ....

A typically decorative map
by Vincenzo Coronelli.
Venice 1691
61 x 46cms 24 x 18in

*An attractive, well-designed
map of this island, little known
to Europeans, by Italy's leading map maker of the period.
A dedication within the title
is made to the Rev. de Fon-
taine of the Society of Jesus
– the Jesuits being the only
Europeans to have insight into
Japan and its society at this
time.*

*The delineation of Japan is
typical of the period with
Hokkaido, here called Yupi,
shown as part of Tartary. A
large, rather westernized
oared barge fills the ocean
between Korea, which itself
had been shown as an island
earlier in the century.*

China and Japan

The earliest appearances on European printed maps of
the Far East were based, for the greater part, on myth
and legend. The few travellers who returned to report
information were not map makers and the distances
travelled increased the unreliability of their estima-
tions. In addition, the currently held belief that the
earth's size was smaller than is actually the case led to
a picture of the Pacific Ocean and Eastern Asia being
much narrower than they really are. For some 60 or 70
years after Magellan, confusion regarding, in particular,

the north Pacific regions and the possibility of a land
bridge between Asia and America, resulted in Japan
being shown at numerous different locations from
mid-Pacific to off the west coast of North America.

Although Benedetto Bordone's *Isolario* of 1528
contained a crude woodblock map entitled *Giapagu*
(Japan), there was no atlas map of the island until
Abraham Ortelius included a map by the Portuguese
Jesuit, Luis Teixeira, in his atlas of 1595.

Eleven years earlier the map of China of another
Portuguese Jesuit, Luis Jorge de Barbuda, had been used
by the same author. These became the standards and

Blaeu subsequently engraved. These 16 maps were beautifully engraved and remained the best maps of the country for almost the next hundred years.

In 1737 Jean Baptiste d'Anville published the next great atlas of China, inevitably based on the Jesuit surveys, and these maps, for many of the regions shown, remained the best available until well into the last century.

Outside mainland China, Korea appeared, until the seventeenth century, as an island and the cartographic development of Japan was somewhat hindered by its self-imposed isolationist policy from 1640.

Unlike China, which had produced its own maps prior to the arrival of Europeans, the early mapping of Japan was provided by the Jesuits. It was not until five years after the exclusion, or at the very least enclosing within the confines of the trading post on Deshima Island outside Nagasaki, of European influence that Japan produced its own maps. Throughout the seventeenth century European mapping of Japan, based primarily on Teixeria's outline, improved only in respect of a more correct orientation of the islands. Internal information was not easily acquired and it was not until the visit of Englebert Kaempfer in 1690 that much new information was added to the map of Japan. The cartographic history of Japan includes some particularly decorative maps and, despite the often retrogressive mapping, provides a fascinating subject for the collector. Ortelius, in his atlas, uses some five different outlines to show Japan, for want of definite information. The various outlines of the northern island, Hokkaido, and its naming, the insular appearance of Korea, the introduction of Japanese letter characters on maps from around 1715 and the decoration, often in oriental style, make for some particularly interesting maps.

were used by subsequent cartographers for many years.

During the next century the European mapping of China has one notable highlight – the *Atlas Sinensis* of Father Martin Martini, published in 1655 by Joan Blaeu as Volume VI of the *Theatrum* and, in due course, as part of Volume XI of the *Atlas Maior*. For many centuries the Chinese had used surveying techniques and the *Mongol Atlas* of Chu Ssu-pen, comprising maps of the Provinces of the Chinese Empire, had been compiled about 1311-12. The Jesuit Father Martini had worked in China from 1637 and produced a translation into Latin of the original Chinese text and maps which

AUSTRALIA AND THE PACIFIC

For the early mapping of the Pacific and, in particular, Australia it is necessary to look at maps of the whole world since knowledge of the southern hemisphere in these regions was predominantly hypothesis until about 200 years ago.

Ptolemy and the other ancient geographers expounded the existence of a great southern landmass simply to balance that on the north. Even the earliest maps, after Ptolemy, of the Ancient World show a land-locked Indian Ocean, its southern coastline being that of this great antipodean landmass. After Magellan's voyage of 1519, which proved that South America, at least, was not attached to this southern land, map makers clung to the ancient principle and extended a supposed coastline from Tierra del Fuego to New Guinea. Typically this 'coastline' would be confidently engraved and a fainter, maybe wavy or shaded line would continue the supposed coastline around the entire

Southern hemisphere. The most northerly points of this theoretical coastline were shown to the south of the East Indies and were frequently marked with the names 'Beach', 'Lucach' or 'Maletur' – names derived from Marco Polo's reports.

At the beginning of the seventeenth century the Dutch made two recorded landfalls in what today is called the Gulf of Carpentaria. Dutch sailors landed in 1605-6 and in 1616 and the process of recorded Australian cartographic history began. Due to the secrecy surrounding many of the charts of these areas the information did not appear in any printed maps until after about 1630 when Henricus Hondius' World

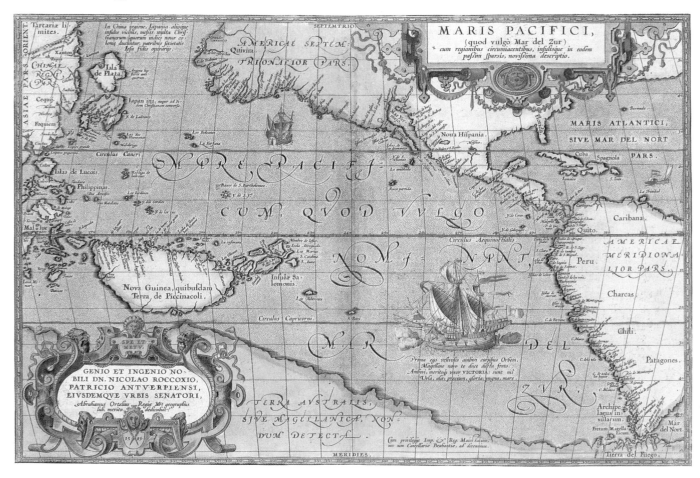

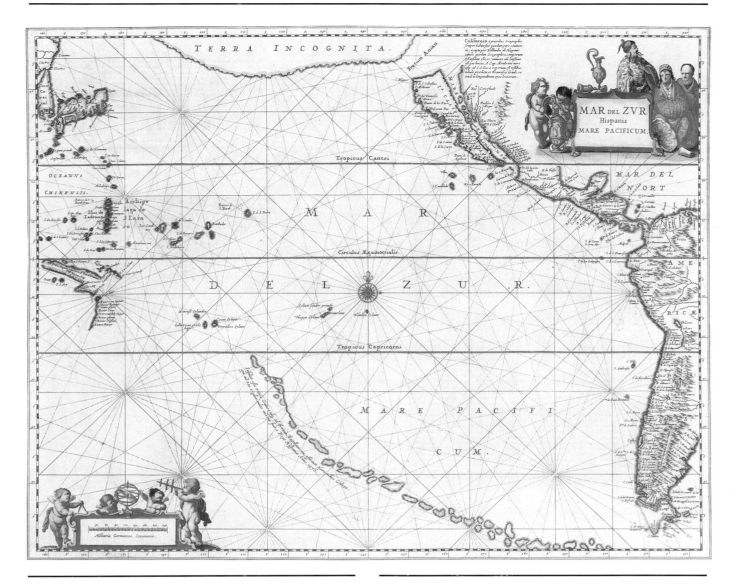

The first printed map of the Pacific
Ocean. From Abraham Ortelius's
Theatrum Orbis Terrarum.
Antwerp 1590
50 x 34cm 20 x 13.5in

*The engraving date, of 1589, appears below the dedicatory
cartouche of this beautiful map. The ocean is shown somewhat
smaller than is actually the case and the west coast of America is
shown in more detail than on any of Ortelius's other maps which
show it. New Guinea is shown as a large island of definite shape
and the great southern continent is also clearly marked. Japan,
in two islands, appears in one of its most bizarre forms.
Magellan's ship – the Victoria – occupies the southern Pacific and
is accompanied by the quotation which translates:*

*'It was I who first circled the World, my sails flying. You,
Magellan, I led to your new found strait, by right am I called
Victoria; mine are the sails and the wings, the prize and the glory,
the struggle and the sea.'*

*The map illustrated is in bright early colour but has some central
staining.*

The Pacific Ocean by Jan Jansson.
Amsterdam c1650
54.5 x 44cm 21.5 x 17.5in

*A decorative map showing a supposed chain of islands in place
of the great southern continent coastline. New Guinea and the
Australian details in the Gulf of Carpenteria reflect the earliest
Dutch landfalls on the new continent. This map was published
after Tasman's voyage of 1642, but his reports did not appear
until over ten years after the event.*

*The decorative oriental objects and figures around the titlepiece
indicate Dutch interest in the Pacific, i.e., a way of reaching the
Orient.*

map first showed these discoveries; in 1636, W Blaeu's
map of south-east Asia extended south to include the
area.

However, the most prominent voyage of this century
was that by Abel Tasman who proved the insularity of
this new continent although his reports were not wholly
acknowledged by mapmakers. Tasman's voyages of

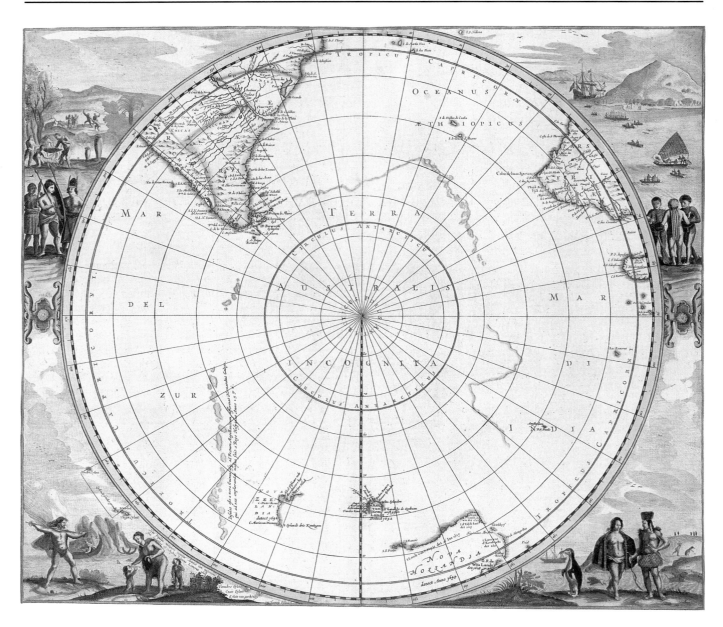

1642-44 were recorded by Joan Blaeu on his World maps of 1646 and 1648 and on many subsequent Dutch publications. However, Nicolas Sanson's World map of 1651 and later issues produced one of the strangest and most confused delineations of any (see page opposite).

The period around 1700 saw some particularly good maps and some particularly bizarre ones. In this latter category the French map makers of the so-called 'theoretical' school of cartography stand out (see page 48). Of particular note for their quality are the globe gores of Coronelli, and the work of de Wit and other Dutch chartmakers who incorporated their details of Australia and Tasman's west coast plotting of New Zealand on the maps of either the East Indies or the Pacific Ocean.

Throughout the seventeenth and eighteenth centuries European knowledge of 'New Holland' was limited to its coastlines since the land appeared barren and

Above
Untitled map of the South Pole

This map summarizes knowledge of the
Southern Seas.
Amsterdam c1657
49 x 43cm 19.5 x 17in

Originally issued by Henricus Hondius in 1641, this copperplate underwent various alterations during its sixty-year life. The first issue has a titlepiece which is changed in 1645 by the addition of Jansson's name in place of Hondius's. In 1657 title and publisher's imprint are removed to make way for the insertion of the coastlines of New Zealand and Tasmania after Tasman's voyage of 1642. Later editions, by de Wit about 1680, and Valck and Schenk 1700, may be found.

It is notable that by this period the uncertainty surrounding the existence of a great southern continent is such that not only are suggested coastlines broken, but also they are indicated by faint lines rather than the confident outlines employed by earlier map makers.

inhospitable, the natives unfriendly, and mosquitos ever present. Many of the landfalls made on the west coast, scarcely reported, were the result of ships being blown off course en route to the Indies. This new land was not encouraging European interest.

The great voyages of exploration of Captain James Cook, only a little more than 200 years ago, were to answer many of the questions still remaining about this hemisphere.

Captain James Cook, born in 1728 at Marton in Yorkshire, had already spent several years around 1760 charting and seeing published his survey of the coasts of Newfoundland. When, in 1768, he left Plymouth on the first of his three voyages to the Pacific and the Southern hemisphere, he could have had little idea that his work over the next ten years would provide more new and correct information about the Pacific than had been acquired since Tasman's voyage of the previous century. Tasman had proved the insularity of Australia but his charting of the west coast of New Zealand still led to the belief in a great southern continent of which that New Zealand coastline was a part.

Cook with a crew of 94 including botanists, artists

and astronomers was ostensibly despatched by the Admiralty to observe the transit of Venus from the island of Tahiti. However, Cook had further secret orders to establish once and for all the extent, if any, of the 'great southern continent'.

Below
A Mapp or Generall Carte of the World

An elegant double hemisphere map by
Richard Blome.
London 1670
53.5 x 39.5cm 21 x 15.5in

This English version of a French World map exhibits contrasting geographical features. The up-to-date nature of much of the Old World and American detail is contrasted by the peculiar Australian and Southern Continent outlines. Blome admitted his plagiarism in this and many other maps – this map by Sanson being published first in Paris in 1651. Nicolas Sanson had used the name 'Beach' derived from reports of Marco Polo to identify Australia. During the seventeenth century there are only a handful of maps which the collector might hope to obtain which pertain specifically to Australia. Accordingly, early World maps are often worth examining for their Australian detail.

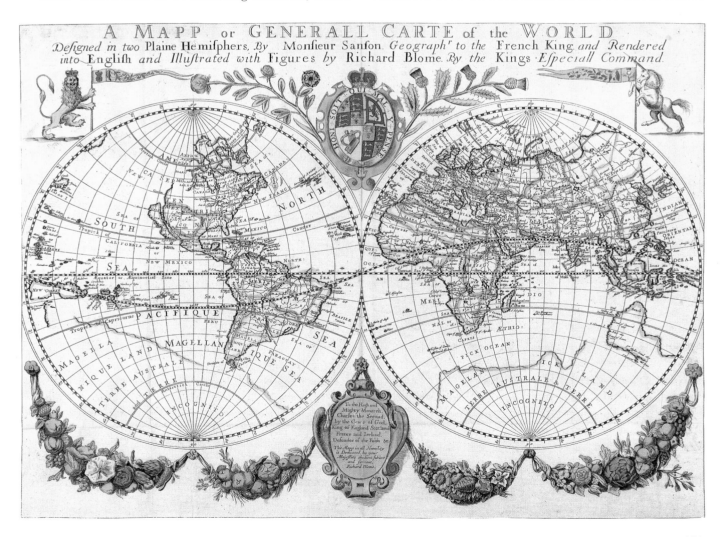

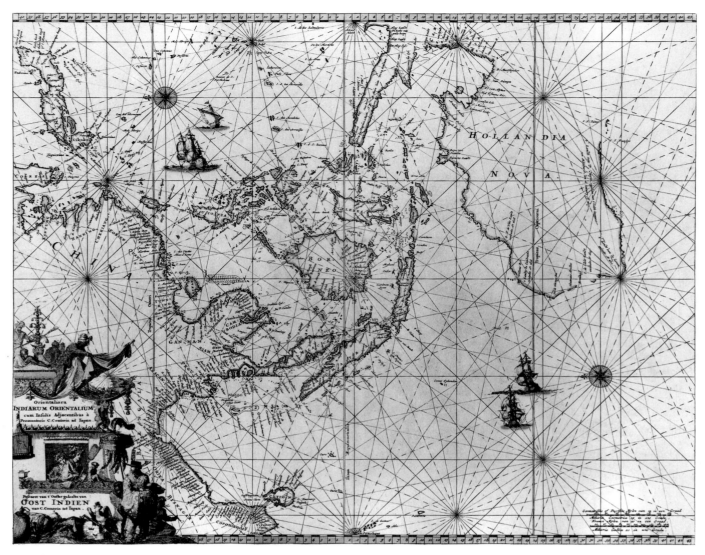

After observing Venus Cook proceeded south and then circumnavigated and charted the New Zealand coasts, thus dispelling any illusions in their attachment to a larger landmass. Proceeding westwards and intending to reach the coast of Van Diemen's Land the *Endeavour* was blown towards the north-west and made landfall on the south-east coast of what we now call Australia. From April to August 1770 Cook sailed and surveyed the east and north-east coasts of the land he called New South Wales, eventually plotting some of the south coast of New Guinea before heading westwards and back home through the Torres Straits, unnavigated by Europeans since Torres himself in 1606.

In his next voyage, from 1772-75, Cook entered the Antarctic Circle and tracked throughout the South Pacific, thereby confirming the non-existence of any large landmass within the South Pacific. His final voyage (commenced in 1776, along the Pacific north-west coast of America, through the Bering Strait, and beyond the Arctic Circle and back south to Hawaii where, in February 1779, he was killed by natives) confirmed the outline of these previously unknown coastlines (see page 31).

It was Cook's second voyage that had seen the introduction of the latest navigating aid – John Harrison's Marine Chronometer – which made feasible the accurate construction of large-scale charts through correct calculation of longitude.

During the next 50 or 60 years after Cook and following the establishment of the English colony at

Right
**Carte Reduite
de L'Australie**

A pre-Cook view of Australia by Robert de Vaugondy.
Paris 1756
29 x 23.5cm 11.5 x 9in

Prior to Captain Cook's voyages there were very few maps of Australia alone, as in this example. This curious map proposes an eastern and south-eastern coastline which joins Tasmania to the mainland, but makes no suggestion for the eastwards extent of New Zealand

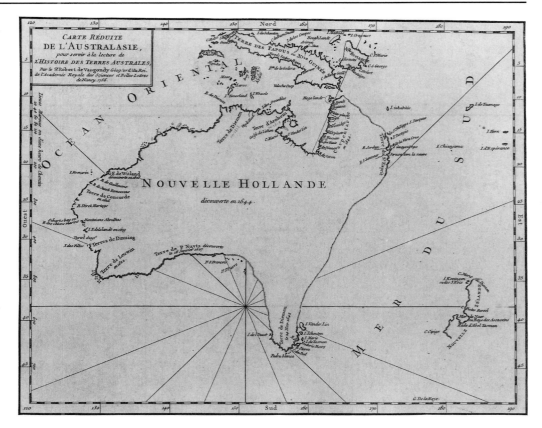

Right
**Carte Reduite
de L'Australie**

A pre-Cook view of Australia by Robert de Vaugondy.
Paris 1756
29 x 23.5cm 11.5 x 9in

Prior to Captain Cook's voyages there were very few maps of Australia alone, as in this example. This curious map proposes an eastern and south-eastern coastline which joins Tasmania to the mainland, but makes no suggestion for the eastwards extent of New Zealand

Sydney Cove in 1788, the cartographic output relevant to the Pacific increased enormously. Besides the maps of major coastlines and islands from volumes of Cook's voyages, there were now specific surveys of the Pacific regions, for example, Bass, Flinders and Freycinet's charts of the Australian coasts, La Perouse and Vancouver in the North Pacific.

Most atlases after about 1800 included a map of 'New Holland', 'Notasia', 'Ulimaroa' (the native name), and subsequently Australia. Also, from this period, detailed maps of sections of Australia became available – probably the best known being those by the Arrowsmiths – Aaron, his sons and his nephew John, who from about 1800 produced over 70 maps devoted to the latest discoveries and explorations in the area.

Contemporary with Cook's reports are versions of his maps published in Italy, France, and of course England. Two of the most attractive of these are by the Italian map publishers, Antonio Zatta and Giovanni Cassini. Prior to Cook there are very few maps which focus on the Australian mainland, Emanuel Bowen's c 1744 version of Thevenot's 1663 map, miniature maps by Herman Moll c 1720, and maps by Bellin in 1753, and Robert de Vaugondy in 1756, are most likely to be found although these are all scarce. A number of other maps may be found from travel books of the seventeenth and eighteenth centuries, as opposed to atlases, but these themselves are all rare. Thus, the early cartographic history of the area is best observed on the more general maps of the period.

THE NEW WORLD

The mapping of the new world is a vast subject in itself and this book can only outline the development of the cartography and survey briefly the great scope for the collector. One of the most exciting elements of the history of new world map making is that the coincidence of the voyages of Columbus, Vespucci and numerous others with the maturing practice of movable type printing, and refinements in woodblock and copper plate engraving and printing, makes it possible to observe and build a collection of the maps which reflect European knowledge of the New World.

At the time of Columbus no printed map extended beyond the confines of the ancient Ptolemaic World – within the next 50 years the world map of Ptolemy expanded in every direction subsequent to the great voyages so that not only did the old world expand to the limits of Africa, Asia and Europe, but also a new world – the other half of the globe – became part of European consciousness.

For simplicity I have subdivided the New World mapping into the following chapter sections: Mapping of North and South America; North America and its parts; West Indies; Bermuda; and Mapping of South America.

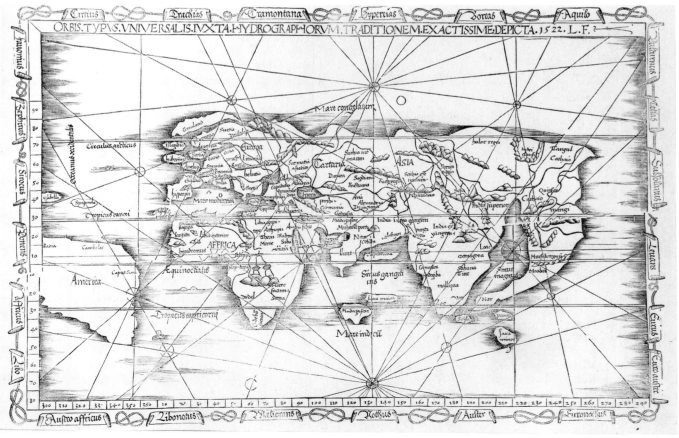

๛ NOVAE INSVLAE XXVI· NOVA TABVLA· ๛

Above
Novae Insulae XXVI Nova Tabula

Sebastian Munster's famous woodblock
map of the New World.
Basle 1552
34 x 22.5cm 13.5 x 9in

*This famous map is found in only one edition with the engraved
latitudinal and longitudinal grid lines, as here. The map shows
North America almost split by the Sea of Verrazano, Japan is
entitled Zipangri, and it is the first to show North and South
America together.*

Left
Tabu Totius Orbis

Laurent Fries' version of Waldseemüller's
World map – the first map from a
Ptolemaic atlas to use the word *America*.
Strassburg 1522
48 x 35cm 19 x 14in

*This unsophisticated woodblock map by Fries shows clearly the
extension of Ptolemaic cartographic theory which accounts for the
misshapen Asian coastline. Africa is shown typically for the
period, though Madagascar and Java in the mid-Indian Ocean
are purely hypothetical. The southern New World landmass is
identified as 'America' and the islands of Isabella (Cuba) and
Spagnola (Santo Domingo) are shown.*

MAPPING OF NORTH AND SOUTH AMERICA

The earliest maps available to collectors now which
show parts of America are the Ruysch World Map (see
page 36), the Sylvanus World Map and the maps by
Martin Waldseemüller (see pages 37 and 139), all
appearing in their respective versions of Ptolemy's
Geographia. After Waldseemüller's unobtainable world
map of 1507, the first printed map using the word
'America' to identify the New World is the rare
woodblock by Peter Apian (1520), based on the 1507
original, which was engraved by Laurent Fries.
Fries himself, in 1522, published a reissue of
Waldseemüller's *Geographia* and, including an unsophisticated World map, used the word 'America' again
to identify the southern landmass of the New World.

Over the next 50 years the expansion of knowledge
of the eastern and south-western coastlines of the
Americas is clearly seen on World maps of the period
and on the few maps of the Americas or parts of the
Americas. Prior to the publication of Abraham
Ortelius's atlas in 1570, the maps by Munster, of North
and South America (see above) and by Gastaldi are
those most likely to be encountered.

The miniature atlas by Giacomo Gastaldi, of 1548, is of particular interest since it includes the first separate map representations of the US north-east coast; the US south and west and Gulf area; South America; and the islands of Cuba and Santo Domingo. There is, however, no general map of the Americas although there are two World maps. In 1561 Girolamo Ruscelli reissued Gastaldi's maps in a slightly enlarged format and added a double hemisphere World map to the atlas. This was the first double hemisphere map to be issued in an atlas and showed North America clearly joined to the Asian mainland by a land bridge, a feature already very much in question.

Sebastian Munster's map of 1540 had already shown North and South America independent from any other land masses. However, many other World maps of this period did not make this distinction. In 1556 Ramusio included a map, attributed to Gastaldi, of the Americas in his *Viaggi ...* . This is the first hemispherical map of the Americas alone and, although the north-west coastline is not clearly defined, there is no apparent land bridge to Asia shown. Ten years after this map Bolognini Zaltieri, in Bologna, issued the map commonly regarded as the first of North America; on this map is the first appearance of the Straits of Annian, possibly proposed by Gastaldi himself, on whose material Zaltieri is believed to have based this map.

The Zaltieri map, issued separately and in Lafreri-style atlases, is rare, but a reduced version of the map,

Above
Universale Della Nuovamente ...

An important map from Ramusio's *Viaggi ....*
Venice 1556
26 x 27cm 10 x 10.5in

The first relatively accurate map of the Western Hemisphere which is still obtainable. Gastaldi, the map maker, has here dispensed with his early depiction of a North Pacific land bridge, but he does not commit himself to an actual coastline. Considering the fact that this map was cut within some sixty years of Columbus's voyages, the map gives a remarkably accurate perception of the two landmasses.

Right
America

John Speed's famous map of the Americas. One of the most recognizable of all old maps.
London 1627-76
51 x 39cm 20 x 15.5in

This important, decorative map was first published in John Speed's 'Prospect of the Most Famous Parts of the World' in 1627. This was the first World Atlas by an Englishman and this is the first map from any atlas to show California as an Island.

The map was engraved by the Dutchman, Abraham Goos, and the border decoration reflects similar 'cartes à figures' produced in Holland earlier in the century (see page 26). Speed's map also reflects Dutch influence in the cartography of South and Central America, but outlines and detail for the north are distinctly different. Henry Briggs' map (see page 162), issued a year prior to the engraving of this – note the engraved date of 1626 – forms the basis for the outline with a little detail left out due to confines of space.

This edition, of 1676, is identified by the publishers' imprint – Bassett and Chiswell – and is one of the few plates of John Speed's atlas to have had improvements or amendments made during its publishing life; further place names were added, particularly within the New England region.

published by Giovanni Magini around 1598 may be found.

In 1570 Abraham Ortelius published his *Theatrum Orbis Terrarum* within which maps of the World and of the Americas are of particular note. Subsequent issues of his atlas include the first detailed map of the South-Eastern section of North America and the first of the Pacific (see pages 143 and 128 respectively).

From around 1600 numerous maps of the Americas appear in world atlases by all the major map makers and

each map exhibits characteristics typical of its period. The following list identifies some of those which may be found and which are worthy of a place in any collection.

1540 Sebastian Munster's first map of the Americas, north and south, as an independent landmass.

1556 The Ramusio/Gastaldi hemispherical map.

1570 Abraham Ortelius's first map of the Americas.

137

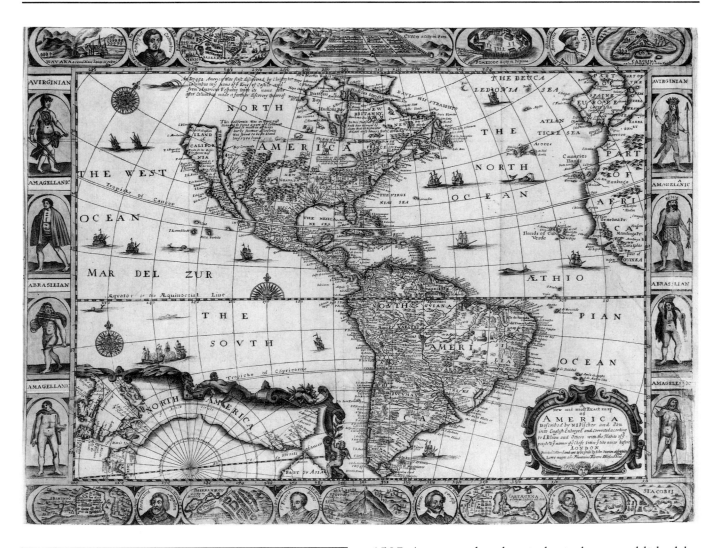

Above
A New and Most Exact Map of America
••••

John Overton's version of a beautiful
earlier Dutch 'carte à figures'.
London 1668
54 x 42.5cm 21.5 x 16.5in

*Published in 1614, a map by Petrus Kaerius provides the
inspiration for the four border strips of decoration around this
map. The Dutch original, engraved by Abraham Goos (who also
engraved John Speed's map) was a finer engraving, but this
version, incorporating up-to-date cartographic theory, has a
charm of its own. Overton credits Visscher and Blaeu, but not
Speed from whom the California Island outline is taken, and
includes, after continental models, two great lakes supplying the
Saint Lawrence.*

1587–8 Abraham Ortelius's re-engraved, improved
map of the Americas.

1595 Gerard Mercator's version of *America* redrawn
and published by his son Michael.

1597 An up-to-date hemispherical map published by
de Bry in the *Grand Voyages*.

1606 Jodocus Hondius's engraving of a new map of the
Americas, famous now for its illustrations of Indians,
at lower left, preparing and consuming Mandioca.

1617 Willem Janszoon Blaeu's engraving of a map
which became famous during the middle years of the
century. Issued in atlases from 1630 onwards, this is
one of the most decorative and sought-after maps of the
Continent (see page 47).

1627 John Speed's world atlas – the first by an
Englishman – which includes his famous map, the first
in an atlas to show California as an Island (see page
137).

1631 Henricus Hondius's re-engraving of an earlier-
published, but rarely issued, map with decorative
borders. This new map is copied by numerous other
map makers.

c 1670 Nicolas Visscher's elegant, new, detailed map

which became the standard for many subsequent map makers.

c 1690 Vincenzo Coronelli's map which, set in a circular surround, was one of the first of North and South America to show the five Great Lakes, i.e., with shorelines defined.

c 1720 Henri Chatelain's magnificent map *Carte tres Curieuse.....* based on a very rare prototype by de Fer of some 20 years earlier.

These are but a few of the maps of the New World. There are many maps available of both the period covered and of more recent times, differing in cartographic merit, size, style and decoration.

Of note at this point are two publications from the early seventeenth century. Published from 1597 to 1615, Cornelius Wytfliet's *Descriptiones Ptolemaicae Augmentum* is regarded as the first atlas of the New World. This contains a World map after Mercator and 18 sectional maps of North and South America derived from Plancius and other printed sources. In 1630 Johannes de Laet's volume *L'Histoire du Nouveau Monde ou Description des Indes Occidentales* included 14 maps engraved by Hessel Gerritz. These fine maps are restrained in style but were the best of the period and influenced the subsequent maps of Blaeu and Jansson, particularly in the regions of the eastern seaboard of North America.

Regionalized maps of areas of the New World were now gaining significance, although with respect to the Continent's interior and north-west reaches, the more general maps were the only ones of the seventeenth century to provide detail in these little-known parts.

NORTH AMERICA AND ITS PARTS

As already stated, the scope for a collector within such a heading is vast, and excellent books devoted to this subject, and to subdivisions within it, already exist. Given the confines of this work I have grouped maps under the following headings for convenience and, hopefully, so that chronological sequences of maps can be followed within specific areas:

Regional maps of North America before 1600.
Maps of North America before 1700.
Maps of Canada.
Maps of the States from 1600:
    New England States to 1776;
    Mid-Atlantic States to 1776;
    South-East States to 1776.
Maps west of the Mississippi before 1800.
Nineteenth-century North American maps.

The reasons for these dates will become apparent and the regions selected will be seen to coincide with the original areas chosen by seventeenth-century map makers.

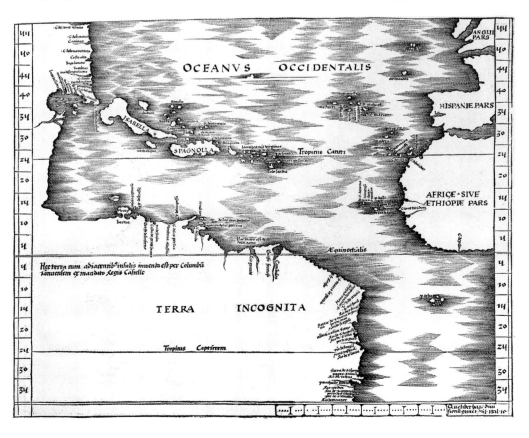

Left
Tabula Terre Nove

Martin Waldseemüller's great map of the Gulf and the known West Indies – the first of any part of the New World.
Strassburg 1513
44.5 x 37cm 17.5 x 14.5in

With its text relating to Columbus, this is the first map which concentrates on those coastlines and islands discovered by Columbus on his three voyages. Waldseemüller reverses his comments of 1507 crediting Vespucci with the discovery of the New World, and his text acknowledging Columbus together with a further reference to an 'Admiral' – the supplier of information for this and the World map (see page 37) – has led to each of these maps acquiring the common title of 'The Admiral's Map'.

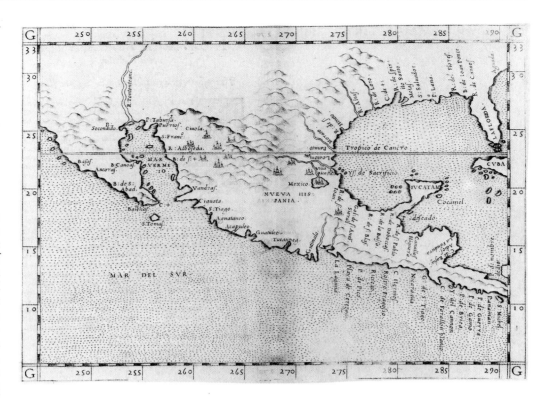

Gastaldi's original 1548 issue of this map is particularly scarce, so this is the earliest relatively obtainable map of the area – the atlas in which this appeared ran to several editions and so maps from it are not difficult to find. Close examination of this plate indicates that, originally, when copied from Gastaldi's map, Yucatan was shown as an island. On this plate traces can be seen of an engraved coastline which has been obliterated and the peninsula has now been correctly delineated. This is an important map showing a more correct west coast cartography than was to be seen a hundred years later when California was shown on most maps to be an island.

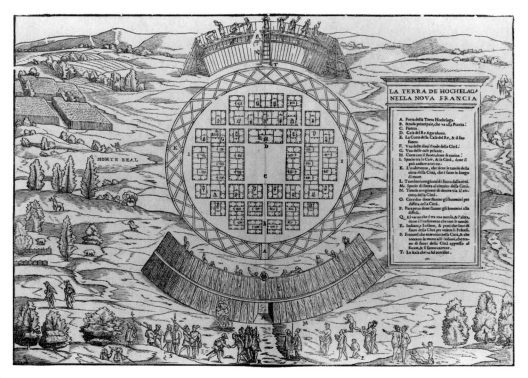

The second voyage of Jacques Cartier, begun in 1535, took up his search for the route to the Indies where his first had left off. Having navigated much of the Saint Lawrence Gulf he proceeded up that river on his second voyage, anticipating a passage to China. After various encounters with native Indians he was brought to the village of Hochelaga, on the site of
present-day Montreal. Cartier named the overlooking hill 'Mont Real' in deference to his King, Francis I, and it is from this that the present name evolves. In this annotated plan can be seen, in the foreground, Huron Indians greeting the Frenchmen.

Regional maps of North America before 1600

The first printed map of any part of the New World is that in Martin Waldseemüller's edition of *Geographia* published in Strassburg in 1513 and 1520. This map is famous as much for its new cartographic depiction as for the statement in Latin appearing on the South American landmass which translates roughly as 'this land and the adjacent islands was found by Columbus

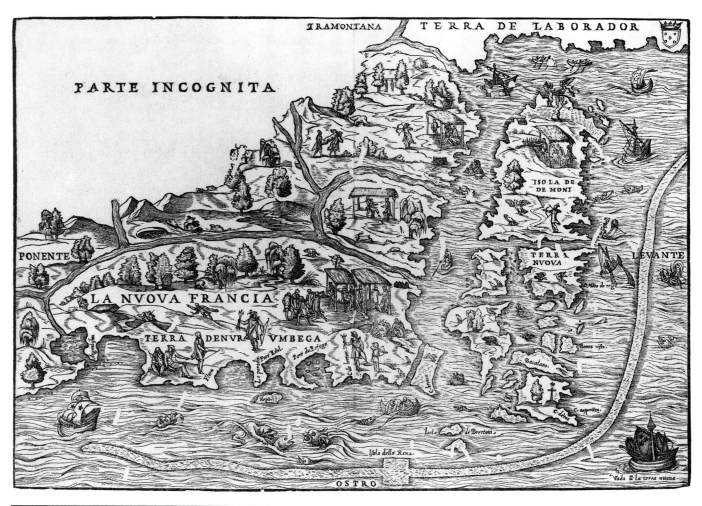

Above
Untitled

La Nuova Francia, Giacomo Gastaldi's
spectacular depiction of the northeastern coastline.
Venice 1556-1606
37 x 27cm 14.5 x 10.5in

This extraordinary map exhibits a licence rarely exercized by engravers of Gastaldi's stature – accordingly some doubt has been expressed over his authorship of this and other similar woodblock maps in Ramusio's work. The map shows the North American coastline from Labrador to present-day New York.

The maze of islands and waterways may respond to the reports of Verrazano who had, about 1524, voyaged in search of a passage to the Indies. Certain locations on the map have been identified against Verrazano's records but the engraving has greater validity as a 'picture' of the country and its inhabitants. This bird's-eye view shows Indians hunting, dancing, fishing and reclining. Fish, in the sea and being hung out to dry, and the island named 'Bacalaos' – Codfish – testify to the seas' natural wealth.

There are three different editions of this map. In the first two editions page numbers 424 and 425 appear in the top corners. The second edition, 1565, is a re-engraved block which, like the example illustrated here, has a monster in the sea between the lower left hand corner ship and the sea dog. Page numbers 353 and 354 appear in the third edition of 1606, which is found in two states – the second showing evidence of termite damage with small areas of print missing (look closely at the sea areas). The example illustrated here is of this second state third edition.

the Genoese under the authority of the King of Castile'. Although it was Waldseemüller who attributed the title 'America' to this land, after Amerigo Vespucci's voyages of 1497, he did not use the word on any maps in his 1513 atlas.

In 1544 Munster's *Cosmographia* was published in Basle and, four years later, Gastaldi's edition of *Geographia* – utilizing text taken from Munster, was issued in Venice with individual maps of parts of the New World. Of the five American maps, two related specifically to the North American mainland and were the first printed maps to concentrate on their respective areas; one covered Baja California, Mexico and the Gulf to Florida, and another the North-East coast from the Carolinas to Labrador. There was only one edition of the atlas but in 1561 Girolamo Ruscelli re-engraved, on a slightly larger scale, these maps which were reissued several times before the end of the century.

Gastaldi's impact on New World mapping was not limited to the series. Ramusio's compendium of travel and voyage reports included maps credited to Gastaldi and related to this exploration of Cabot, Verrazano, Cartier and others in the Americas.

The 1584 edition of Ortelius's great atlas saw the inclusion of a map of Spanish Provinces in the Americas and, five years later, a map of the Pacific. This was the first map of the great ocean and showed the west coast of the Americas in good detail (see pages 143 and 128).

Before the end of the 1500s, in fact during the last decade, a series known simply now as 'de Bry the *Grands Voyages*' was published in Frankfurt-am-Main in Germany. This sequence of publications includes a number of maps whose influence extended well into the next century. Of most importance are the maps of Virginia centred on Raleigh's Roanoke Colony, issued in 1590, and a year later the maps of Florida and the southern states after Jacques Le Moyne, a respected French artist who accompanied Landnonnière's expedition to Florida in 1564. A further map, of 1594, is the first detailed one of the West Indies and adjacent coastlines and is one of the most decorative to show the area. In 1597 a map of North and South America incorporating much of the information already expounded within the *Grands Voyages* was published; this, accordingly, was one of the best maps then available showing the two landmasses.

Finally, Wytfliet's atlas, of 1597, includes maps of individual areas although these, in the main, are taken from larger maps rather than being original as were the earlier maps mentioned.

Maps of North America before 1700

A comparison of the maps illustrated here by Jansson of *c* 1640 (see page 144) and the late

issue by Elwe of Jaillot's plate (see page 149) indicates the rapidly changing and improving cartography of North America. Prior to 1625 when Henry Briggs' map (see page 160) was published, there were very few maps of North America independent of South America.

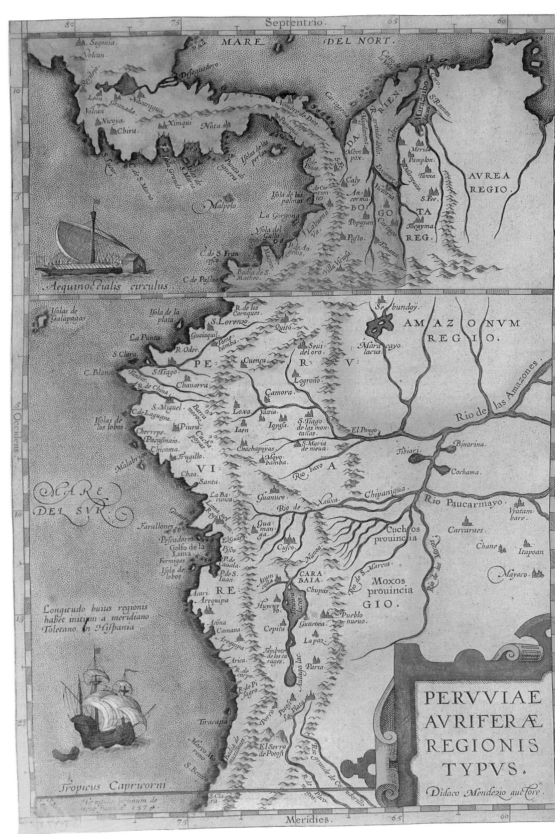

These, as the Briggs map shows, contained virtually no inland detail except along a few rivers and, lacking in the Briggs maps, some detail in the southern and western states emanating from the Spanish missions in those regions.

With the publication of the Briggs map the insular California theory was promulgated and this appeared on most maps of the region until the early decades of the next century.

The list below identifies certain seventeenth-century maps and their significant features in the development of North American cartography.

1625 Henry Briggs' *The North Part of America* was published in London. This was not only the first map to show California as an Island but also the first to name Hudson's Bay and the first of North America in English.

1636 Jan Jansson's *America Septentrionalis* was the first map of North America showing California Island to appear in a Dutch Atlas. The Saint Lawrence is shown, typically, flowing from one Great Lake.

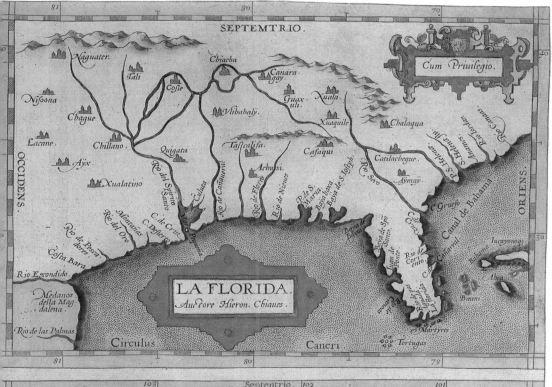

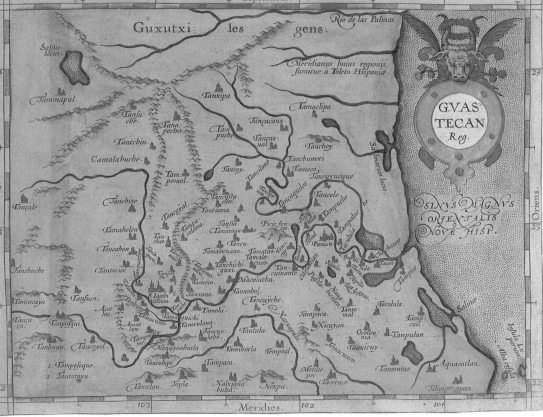

Left

La Florida, Peruviae and Guastecan

Three maps of the Spanish Provinces in the Americas by Abraham Ortelius. Antwerp 1584
46 x 35.5cm 18 x 14in

In fine original colour, these three maps, on one copperplate, were published in all editions of the 'Theatrum Orbis Terrarum' after 1584. The map of Florida is of particular interest here since its content, based on Spanish reports from de Soto and others, became the standard representation of the area on many subsequent maps.

1650 Nicolas Sanson's *Amerique Septentrionale*. This map by one of the leading figures in French cartography shows new detail within the New Mexico region and is the first to indicate the existence and location of all five Great Lakes.

1669 Richard Blome's *A New Mapp of America Septentionale* is a straight copy of Sanson's earlier map and as such is the first English map to show the five Great Lakes

1674 Alexis Hubert Jaillot's *Amerique Septentrionale* is also a copy of Sanson's cartography but utilizes an irregular, inundated north coast delineation of the California Island, adopted by Sanson for the second edition of his 1650 map issued in 1669. Maps from this series are famed mostly for their spectacular presentation, although this map adds new detail of French Missions around the Lakes.

1696 Vincenzo Coronelli's magnificent two-sheet *America Settentrionale* is the largest-scale map of North America of the period. Coronelli's large maps of the continents are all spectacular cartographic items and this map, with its numerous vignette illustrations taken from de Bry, its large titlepiece and decorative embellishments, is, perhaps, the most impressive.

A large California Island is prominent, the Great Lakes are shown more accurately than on any other

Below
America Septentrionalis

The first Dutch atlas map of North America to show California as an island, by Jan Jansson.
Amsterdam c1640
56 x 47cm 22 x 18.5in

This finely engraved map appeared in the Jansson /Hondius series of atlases from around 1636 and is one of the most elegant maps of North America. Note the charming vignette views of native animals and birds. The particular example shown is in fine early colour, heightened with gold.

map of the period, but the Mississippi is placed too far West. Coronelli, with his French associations, had access to the reports of, amongst others, Marquette (1673) and La Salle (1682), and his map and globe production at the end of the seventeenth century forms a highpoint in cartographic history.

1700 Guillaume de L'Isle's *L'Amerique Septentrionale* indicated a new phase in North American cartography. Although not generally accepted, the theory of California as an island had been disproved by the crossing of Father Eusebio Kino from New Mexico to the Pacific Ocean two years earlier, and maps from this time following de L'Isle's lead, began to show, if not a Californian peninsula, at least a doubt about the coastline.

It was not until 1747 that California was officially proclaimed 'not an island' by King Ferdinand VII and by this time a great number of maps had been published maintaining the island theory against the indications of Father Kino. Map makers of all nationalities were guilty of perpetrating this myth, none more so than Herman Moll who stated about 1720 that he knew English mariners who had sailed around the island! From Italy the maps of Rossi and Petrini; from France, Sanson, Duval and Nolin; from Germany, Scherer, Homann and Seutter; from Holland, Valk, Van der Aa and Van Keulen; from England, Seller, Overton, Wells and Moll – to mention but a few – can all be found, each showing a different style and delineation for a country whose bounds were constantly questioned and advanced.

As this knowledge developed so, from the atlases of Wytfliet and de Laet, and from other, single identifiable 'mother' maps of certain areas the cartographic development of North America can be seen.

Maps of Canada

The cartography of Canada is, of course, linked absolutely with that of North America and despite the maps which incorporate the name 'Canada' within their title, any study of maps of Canada will involve maps of the wider areas of North America, the North Pole, the North Atlantic or Pacific Oceans. Canada as it is known today is a landmass of which, even less than one hundred years ago, vast stretches were unknown. Seventeenth and eighteenth century cartography shows advances along two particular fronts – the search for a north-west passage to the Indies led along a northerly track towards the Davis Straits, and along the southerly track up the Saint Lawrence and into the heart of the North American continent. It is along these two lines that the basic early cartographic developments may be seen.

In 1497 Cabot sailed from England for the North

Above
Untitled Globe Gore

Engraved for one of Vincenzo Coronelli's
great globes.
Venice c1690
20 x 35.5cm 8 x 14in (approx)

This fascinating engraving, designed to be pasted onto a globe, shows Coronelli's correctness in the delineation of the Great Lakes and his error in locating the north and lower reaches of the Mississippi too far westward. Arctic Canada is well depicted and the numerous vignette scenes refer back to illustrations from de Bry's volumes.

West Passage; in 1524 Verrazano, and in 1534 Cartier sailed the coastlines, and the latter the Saint Lawrence, to plot the eastern Canadian coastline as used by Ortelius in 1570. However, the greatest name in early Canadian cartography is that of Samuel de Champlain whose map of 1613 is one of the most important, and most rare, of all American cartographic pieces. This map extends from Newfoundland westwards and has a finely engraved lower border of native portraits and

The North West Passage

During the second half of the sixteenth century English and Dutch voyages in search of a north-east passage from Western Europe to the Orient by way of the Scandinavian and north Russian coasts had been thwarted by natural conditions. Equally, the search for the north-west passage through, or over, North America to the Pacific had encountered Arctic ice or the huge landmasses of the Americas.

The voyages westwards, although not attaining their intended goal, had reaped substantial rewards, and European settlements were becoming established along the coasts south of the St Lawrence Gulf from the early 1600s. The St Lawrence, even into the seventeenth century, was seen as the way to the west – possibly through the fabled sea of Verrazano. However, English voyagers persevered in the search for a more northerly route and the names of Frobisher, who sailed in 1576-8, Davis 1585-1587, Hudson 1607-1610, Button 1612-13, Baffin 1615 and others are all recorded on maps of the period.

Inevitably, by the mid-seventeenth century, interest and financial and physical support for the search could not be maintained in the light of more immediate colonial acquisition on the North American mainland, and, although throughout the eighteenth century, some English and many French maps and charts suggested channels from the north Pacific to northern Canada and Hudson Bay, the outline of the greater part of the Canadian north had been established.

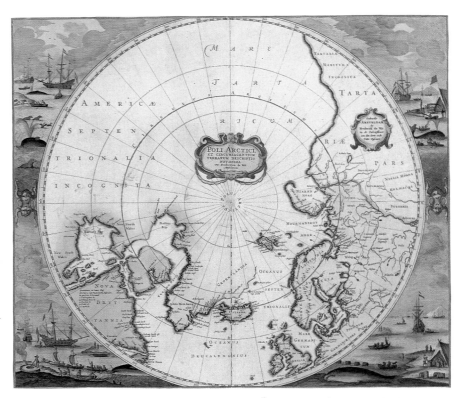

Poli Arctici

Frederic de Wit's decorative map of the North Pole.
Amsterdam 1675
49 x 43cm 19 x 17in

This attractive polar planisphere illustrates clearly the state of knowledge of the northern seas as far east as Nova Zembla. Different stages and processes of the whaling industry are shown.

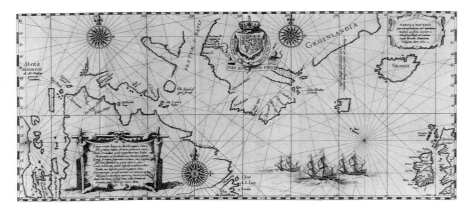

Tabula Nautica ...

Henry Hudson's last voyage, finely engraved by Hessel Gerritsz.
Amsterdam 1612
52 x 22cm 20.5 x 8.5in

This beautifully engraved chart is better known in the reduced version which appears in Part X of de Bry's 'Petits Voyages' of a year later. This original issue appeared in the pamphlet published by Gerritsz entitled 'Detectionis Freti ...'. Henry Hudson had sailed for the Dutch in search of the north-east passage in 1607 and a year later to the coast of North America, where exploring the New England coast, he found and negotiated inland a distance of the river which bears his name.

On his return to England Hudson was contracted by London merchants to search once again and it was on this voyage that he passed the Davis Straits and into the Bay which now also bears his name. As a consequence of the harsh conditions and, in particular, of their captain's treatment of them, his crew mutinied and left Hudson marooned, but their reports – possibly used as barter against their executions provided the information for this important map.

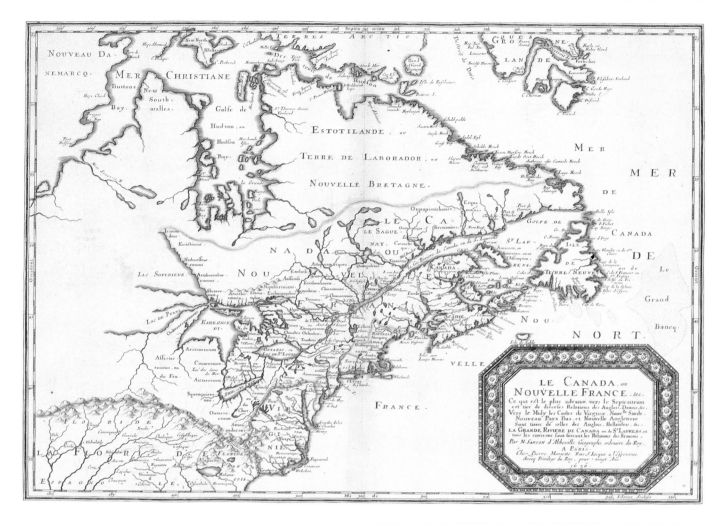

Above
Le Canada ou Nouvelle France

Nicolas Sanson's famous map – the first
of the area to show five Great Lakes.
Paris 1656
53 x 39cm 21 x 15.5in

*An important and uncommon map combining Champlain's
geography around the Saint Lawrence and Henry Hudson's in
the Arctic regions. This example is in early outline colour.*

diagrams of herbs and plants. This was surpassed, though, by subsequent maps, including that more frequently found by Pierre Du Val, dated 1653, which is in fact a reprinting of a map engraved in 1616. This, unlike its predecessor, includes the Canadian Arctic coasts.

The cartography of Canada during the seventeenth century can be seen in the previously mentioned general maps and also in the increasing number of chart books concentrating on the northern coasts. In 1650 Nicolas Sanson published his map of North America, the first to indicate the five Great Lakes, and in 1656 his map *Le Canada* became the first to show this feature in detail. Including all of the known north coast and extending as far south as the Carolinas, this map was copied on a reduced scale for Sanson's quarto version of his atlas, published in several formats between 1682 and 1734.

Guillaume de L'Isle's map of 1703 shows the next accurate Canadian detail – the five Great Lakes are accurately shown, as are the areas to the north and to the west where, however, the mythical 'Rivière du Longue' is sited.

From this date, the collector can find more detailed maps of specific regions. With exception of localized charts from sea atlases, the maps of the mid-century by and based on Bellin provided the best description of the Lakes and the Saint Lawrence hinterland then available. Subsequently, over the next 20 or 30 years and as a result of the hostilities between the British and the French, a large number of detailed charts and maps were engraved and published showing the latest detail of many parts of the Americas. Most of these were issued by London publishers, Thomas Jefferys, Robert Sayer and John Bennett, and also in topical magazines.

By the early nineteenth century, detailed maps of Upper and Lower Canada were being published by Arrowsmith, Wyld and many others along with details of the northern parts and the relatively newly-

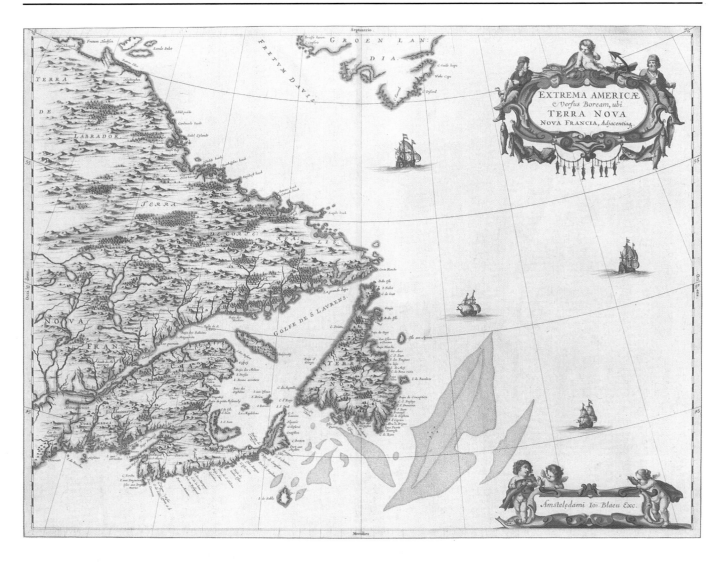

discovered coastlines of the great North West.

Northwestern maps show, in the early days, great conjecture regarding a coastline which was not, on record, visited until Cook on his last voyage in 1778. Cook voyaged from about 44.5 degrees North on the North American west coast and travelled into the Arctic Circle through the Bering Straits. Later on this trip he was murdered but his navigational skills provided the outline of much subsequent surveying work, soon to be followed by Vancouver and La Perouse.

The nineteenth century sees some fine maps, not all of large-scale, repeating the work of Arrowsmith and others. For the collector there is a good quantity available although the maps still tend to show the whole of Canada or sections along the Saint Lawrence.

Newfoundland is well represented throughout cartographic history owing in part to its 'barrier' location at the mouth of the Saint Lawrence and also because of its natural harvest of the seas which interested European sailors from the outset. Beyond the Ramusio map of the North-East (see page 141), there are maps in Bertius's miniature atlas of the island alone from as early as c 1610, some of which are particularly decorative.

Above
Extrema Americae

A beautifully designed map from Joan Blaeu's *Atlas Maior*.
Amsterdam 1662
56 x 44.5cm 22 x 17.5in

Taking much of the detail from Champlain, this is the only map in the atlas to detail Canada apart from that of the whole of North and South America. Newfoundland is central to the map and its adjacent sandbanks are clearly marked. The title cartouche is unequivocal in its indication as to the produce of the area – Fisherman are seen with their nets and numerous codfish. The map illustrated is in fine original colour.

Maps of the States from 1600

From the start of the seventeenth century, it is possible to identify the maps of certain map makers which set the subsequent pattern for the cartographic development of the area that became the United States.

Maps in Theodore de Bry's *Grands Voyages* provide

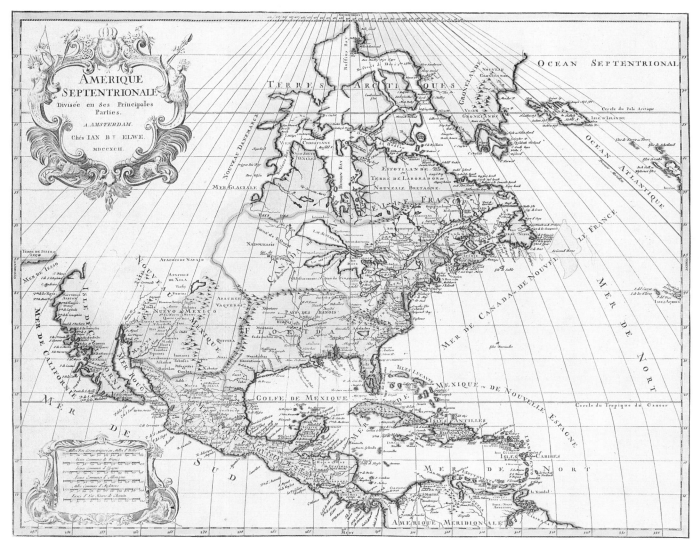

Above
Amerique Septentrionale

Published by Johann Baptist Elwe, this
shows a standard eighteenth-century
delineation of North America.
Amsterdam 1679-1792
58 x 47cm 23 x 18.5in

*The copperplate of this map was originally produced and published
by Alexis Hubert Jaillot over one hundred years earlier. During
this time the original island, California, has been improved to a
more correct peninsula and the five Great Lakes have been
completed whereas the original was undecided. The east coast is
still, however, outdated. This example is in typical delicate original
colour.*

the first indications as to regions which would appear
most frequently as separate maps. In 1590 and 1591 the
maps of Virginia by John White and Florida after
Jacques Le Moyne were published. These provided
much of the detail for Jodocus Hondius' map *Virginiae
Item et Floridae ...*' of 1606, and subsequent maps by,
amongst others, Blaeu (1640) and Ogilby (1670).

Captain John Smith's map of 1612 established the
standard seventeenth-century format for the Chesapeake regions and few subsequent atlases which
included regional maps of the States were without a
map of Virginia and Maryland.

For the cartographic development of the areas further
to the north of Cape Charles, it is necessary to look at
Jan Jansson's maps of 1636 and 1651. In 1620 the
Pilgrim Fathers at New Plymouth had established the
first permanent settlement in New England and, three
years later, the Dutch Colony of New Netherland was
created. In 1630 Hessel Gerritsz' map of the Eastern
Seaboard was issued in Joannes de Laet's book on the
New World. The ascendancy of Dutch interest and
cartographic prowess in the area has fortunately left us
with some beautiful and important maps. It is

appropriate, accordingly, to look at the cartography of
each of these areas, in its respective grouping.

New England States to 1776

For 140 years European experience of North America
extended inwards from the coastlines so that, by the
end of that period, all of the country between the

Atlantic and Mississippi was relatively well-known.

In 1635 Guillaume Blaeu's *Nova Belgica et Anglia Nova* was published – the first map to concentrate on the New England area and one of the most attractive early maps of the area. The map has inset views of native life, animals and birds and is orientated with North to the right of the sheet. Blaeu's map appeared in subsequent issues of the Blaeu' atlases for about the next 50 years, but it lacked the importance and influence of Jansson's map of one year later.

Jansson's 1636 map was followed in 1651 by one which was to instigate a whole group of successors which dominated the area's cartography for the next 100 years. Jansson's *Belgii Novi* was, when published, the best map of the area, showing English, Dutch and Swedish settlements along the coasts and rivers. Additionally, as in the Blaeu map, views of •Indian native life and vignette scenes of animals and birds occupy the blank areas. In 1655 Nicolas Visscher redrew and re-engraved the map with the addition of a few place names and minor alterations to vignette decoration, but particularly with the addition of a view, at lower right, of *Nieuw Amsterdam*, i.e., New York (see page 24). This exceptionally decorative map of an area which held as much interest to Europeans at that time as it does now, spawned copies and variations by Justus Danckerts (1655), by Adrien van der Donck (1656), by Hugo Allard (1656), later issued by Ottens in two editions and by Matthias Seutter (1730), later issued by Lotter and Probst. Versions of this map which lack the view were issued by John Ogilby (c 1670), by Francis Lamb for Speed's *Prospect* ... (1676), and by Pieter Vand der Aa who altered and reissued the Ogilby plate in 1714. All maps from the series are particularly sought by collectors.

By about 1700 in England, map makers such as Morden and Moll and the chart makers Seller, Thornton and Lea, were producing maps from original sources and these rapidly improved the rather staid cartographic picture which had been maintained throughout the latter part of the 1600s. Equally, the Dutch chart makers – Van Keulen in particular – were

Right
Belgii Novi

The second state of Jan Jansson's 'mother' map of the northeastern states.
Amsterdam 1666
52 x 44.5cm 20.5 x 17.5in

The one earlier issue of this map lacks the dedicatory cartouche to Gualthero de Raet, knighted by Charles II in 1660. Even without the view of New Amsterdam which was added by Visscher (see page 24), this is a most distinctive and elegantly designed map. This copy is in early outline colour.

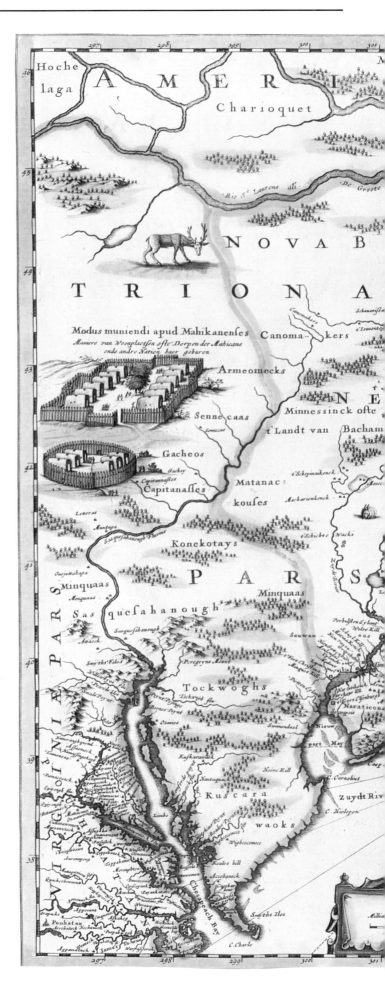

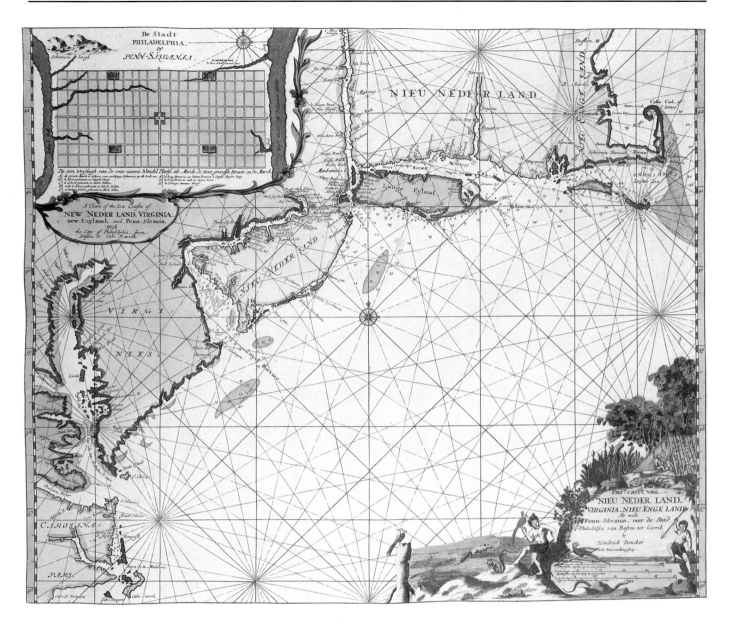

Above
**New Neder Land, Virginia ... Nie Neder
Land ...**

A fine, rare chart by Hendrick Donker.
Amsterdam 1688
58.5 x 50.5cm 23 x 20in

*This magnificent chart has, not unusually, a double title in both
Dutch and, in this case, English. Besides this chart's rarity, it is
particularly unusual in incorporating a town plan within an inset.
Philadelphia was only laid out in 1682 so this is, therefore, one
of the earliest plans of that city. The chart is in fine original colour
and was issued in 'Zee Atlas ofte Waterwerelt ...'.*

producing newly-surveyed charts to assist the ever-
increasing trade within the area.

Within the general context of maps prior to 1776,
there are two major works of note relating to the eastern
half of North America. In 1733 Henry Popple issued

his map of *The British Empire in America ...*, a
twenty-sheet map, measuring overall about 240 x 230
cm, extending south from Hudson Bay and Labrador
to Latin America, and including New Mexico and the
Midwest eastwards. This was the first large-scale map
of the country and includes inset details of many of the
New World ports, harbours and islands. When found
complete this is, of course, a magnificent item.
However, individual sheets are sometimes seen and
these provide great opportunities for the collector to
obtain larger-scale representations of his particular area
than are otherwise available. The second, and more
important of these maps is John Mitchell's of 1755.

Mitchell's *Map of the British and French Dominions in
North America ...*, which measures some 196 x 137
cm, has been regarded as the most important map in
American cartographic history. Needless to say, examples very rarely appear on the market. Mitchell was a
man of many interests, being a medic, biologist,

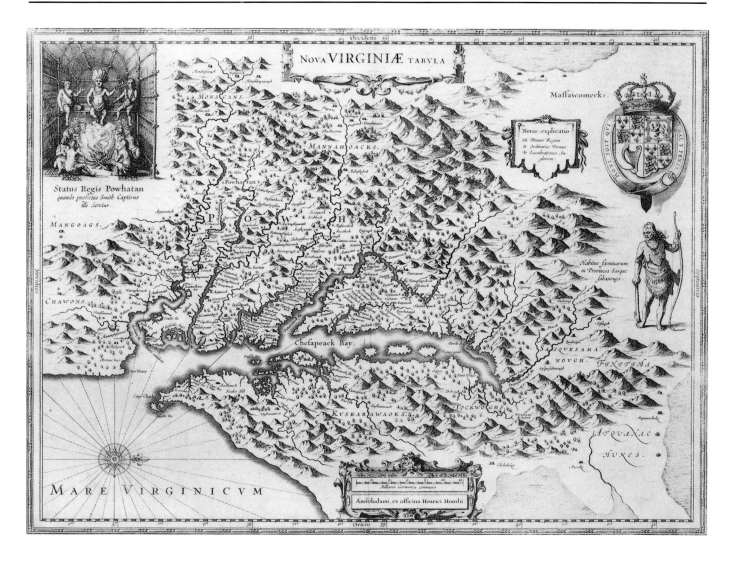

Nova Virginiae Tabula

Henry Hondius's version of Captain John
Smith's famous influential map.
Amsterdam 1633
49 x 38cm 19.5 x 15in

*Captain John Smith's map is here copied by Henricus Hondius.
The illustration of an Indian warrior and the internal scene of a
native hut are also taken from the Smith original. Smith was one
of the first settlers and produced a remarkably accurate map.*

chemist, botanist and map maker. This, however, is
his only known map, but one which finds an important
place in American history. Maps of this time,
particularly in such nationally sensitive areas as this,
served a geographical and political purpose, defining
or merely claiming boundaries and, effectively, putting
on paper territorial claims. This English map disputed
French claims in North America and was actually used
(in its third edition) by the negotiators at the 1783
Treaty of Paris.

European rivalries over the fate of the American
colonies inspired, around the mid-eighteenth century,
some fine, detailed cartography produced by the most
up-to-date techniques and with the urgency of pressing
military requirements.

Mid-Atlantic States to 1776

Captain John Smith's map of Chesapeake Bay and the
surrounding country proved to be a durable piece. It
was reprinted in numerous editions of de Bry,
Hondius/Jansson, Blaeu, Ogilby/Dapper, Van der Aa
and Speed publications throughout the seventeenth
and early eighteenth centuries, even after becoming
outdated with the publication, in 1673, of Augustine
Herman's new survey of the area.

In 1670 John Ogilby published his *America*, based
on the work published earlier that year by Arnold
Montanus but included his own reports of the newly
settled colonies. The book, with maps of all parts of
the Americas, North and South, was the best
compendium of information on the New World of its
time and included, in later issues, *Nova Terrae-Mariae
Tabula*, the first available map of Maryland (see page
154).

John Speed died in 1629 but publication of his world atlas *The Prospect of the Most Famous Parts of the World* was maintained and, in 1676, Francis Lamb engraved a map incorporating Smith and Herman's detail to produce one of the most desirable collectors' maps of the region entitled *A Map of Virginia and Maryland*.

Herman's large-scale map is known now in only a few copies but its influence is recorded on maps well into the next century. After the 'Speed', John Thornton and Robert Greene (*c* 1678), Robert Morden (1680) and Christopher Browne (1685) published maps based on Herman's map. The Thornton-Greene map established the area's mapping on a north-to-south vertical, as opposed to horizontal, axis as shown in the original Smith map. This format was followed in land maps, though not in charts, by most subsequent map makers – Moll's map from 1708 is found in a number of different editions and Emanuel Bowen's decorative map of 1747 is particularly popular. With regard to charts of the area, that by John Thornton and William Fisher of 1689 (and many later editions published by Mount and Page), and one by Pierre Mortier of 1696, are the most regularly encountered.

By the mid-eighteenth century, in the wake of Popple's map of 1733, the requirement for large-scale detailed maps of all areas along the North American Atlantic coast, particularly for use by the British authorities in their struggles with the French for control

---

Below
Nova Terrae-Mariae Tabula

The first obtainable map naming and defining the present state of Maryland – from Ogilby's *America*.
London 1671
38 x 29cm 15 x 11.5in

In September 1635 the first printing of Maryland's charter was published accompanied by the predecessor of this particular map. This is virtually unobtainable. This map, though not very accurate, was to become especially important when the Penn family of Pennsylvania claimed that the bounds of Pennsylvania should be further south, encroaching on Maryland, than had originally been granted. This cartographic error was corrected (though the matter was by no means finished) by John Ogilby when he issued this version in later editions of his book on 'America'.

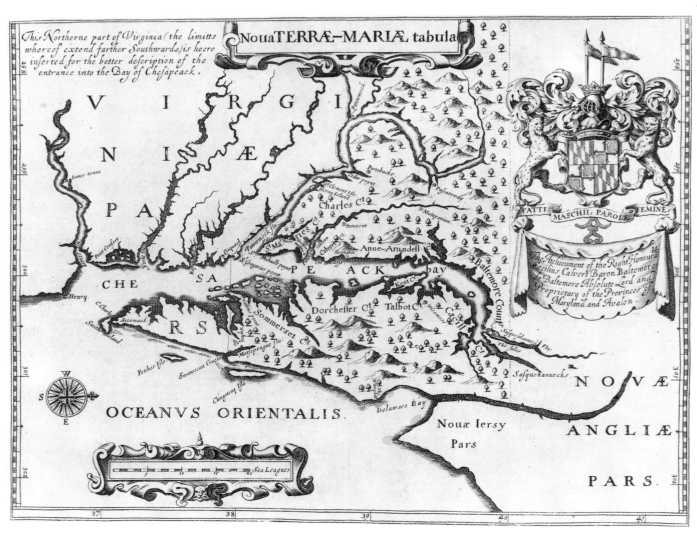

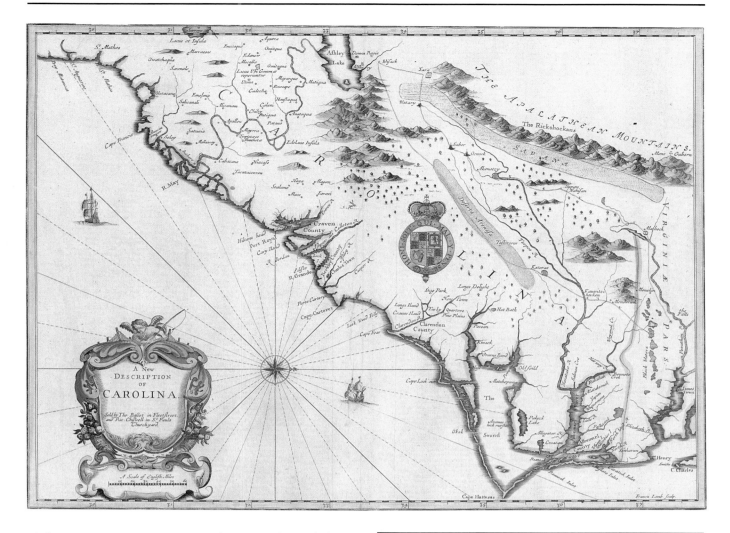

of this new Empire, necessitated new work, and from around 1750 for the next 30 years some great maps were published.

Of particular relevance to Virginia and Maryland was the large-scale map of Joshua Fry and Peter Jefferson first issued in 1754 and, in later editions around 1776. This shows much internal detail and indicates for the first time the correct direction of the Appalachian ranges. Most copies of this map to appear on the market were published in *The American Atlas* (editions of 1776 and 1778) by Thomas Jeffreys, but other, separate editions by Laurie and Whittle were recorded, along with a later edition of 1794. In France copies of the map were issued by George Le Rouge in 1778, a reduced version having been published in the Robert de Vaugondy's *Atlas Universal* after 1755.

Small, but attractive maps can be found from publications such as *The Gentleman's, Political* and *Universal* magazines. Very popular in this period of Colonial expansion, maps from these volumes are often amongst the earliest maps of the individual states available on the market and should be relatively inexpensive. Because of the speed with which they were produced, many of these maps had a topicality which atlas maps lacked and which is only matched in

Above
A New Description of Carolina

A decorative map engraved by Francis
Lamb for Speed's *Prospect*.
London 1676
52 x 38cm 20.5 x 15in

A sought-after map published by Basset and Chiswell in only one edition of Speed's world atlas. The map is based on the First Lords' Proprietors Map found in some issues of Ogilby's 'America'. Like other maps in this series this map has text on the reverse describing the geography, history and development of the area – there is a particularly detailed account of John Lederer's expedition of the early 1670s which provided much information on the interior cartography of the Carolinas.

the much more rare broadsheet maps showing events of the war in North America which are sometimes found.

South-East States to 1776

The maps of Florida by Ortelius, first published in 1584, of the Virginian coast around Roanoke by John White, published by de Bry in 1590, and of Florida and the South East coast by Jacques Le Moyne, also by de Bry in 1591, formed the foundations of south-eastern cartography. The map by Ortelius is not too hard to find, having been reprinted about 25 times, but the other maps, finely engraved, are rare and particularly sought after. These latter maps formed the basis for Jodocus Hondius' map of 1606, engraved for his edition of Mercator's *Atlas*. This map also incorporated vignette views taken from the illustrations used by de Bry in his *Grands Voyages*.

For over 60 years the cartographic depiction of this area varied little – maps by de Laet (1630), Blaeu (1640), Sanson (1656), and the Dudley charts (1647) added only small amounts of detail to the earlier outline. In 1670, the book *America* by Arnoldus Montanus and John Ogilby, in German and English editions respectively, included in early issues a map taken directly from Blaeu. This was later substituted by a map now called *The First Lords' Proprietors' Map* in deference to the Governors of Carolina which deviated from the mould of the earlier maps and, with *The Second Lords' Proprietors' Map* of 1682 by Joel Gascoyne, set the direction for subsequent mapping of the area. A detailed and influential map of c 1685 by John Thornton and Robert Morden of South Carolina

was copied by Dutch and English map makers including Mortier (see page 20) and Moll.

At this point (i.e., c 1700), few maps concentrated in detail on the area of Florida or the areas westwards, whereas the East coast was better served. Needless to say, little was known of these areas although a map by Emanuel Bowen (*A New Map of Georgia*, 1748) shows at a good scale the area of the south-east states from the Mississippi to Savannah. Identifying Indian vil-

---

Below
Barre et Port de Charles-Town

A French copy by Le Rouge of a detailed plan, issued originally in London by Sayer and Bennett in 1776.
Paris 1778
61 x 45cm 24 x 18.5in

This detailed chart is typical of the large-scale, detailed plans produced as separately-issued broadsheets depicting events of the War in America. Sayer and Bennett, in London, were the major producers of these now rare maps.

This map describes the disastrous attempted assault on Charleston by British forces in mid-1776.

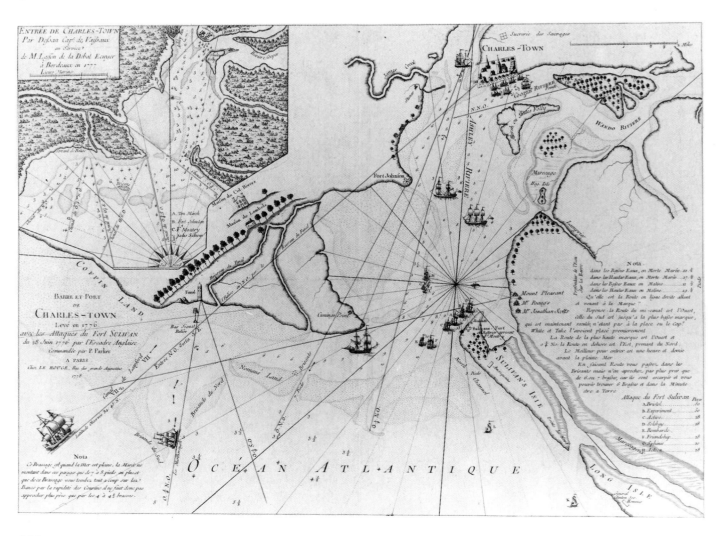

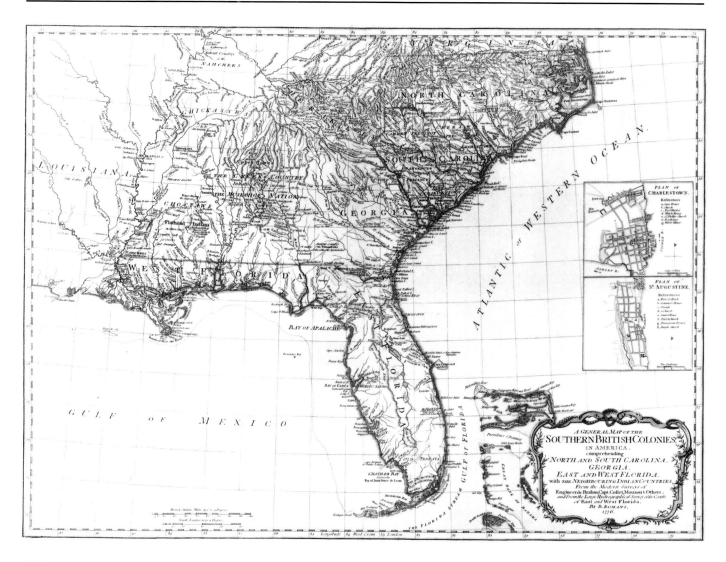

Above
A General Map of the Southern British Colonies

The first edition of this detailed map published by Robert Sayer and John Bennett.
London 1776
64 x 50.5cm 25 x 20in

This standard representation of the area at the commencement of the Revolution was issued first in the American Military Pocket Atlas' – the so-called 'Holster' Atlas which included five other maps of the east coast. As the title continues '… being an approved collection of correct maps … of British Colonies, especially those which now are or probably may be the theatre of war'. These were the best general maps available of the period and were folded into pocket size to fit a holster or pouch for use in the field.

The map was later issued and is more frequently found in Kitchen's 'A New Universal Atlas' with title and imprint changed and published by Laurie and Whittle.

lages, cross-country routes and so on, this map occasionally can be found on the market. Covering larger areas, the maps by Guillaume de L'Isle are of importance and influence, *Carte du Mexique et de la Floride …* of 1703 covers the greater part of today's United States, Central America and the West Indies, and *Carte de la Louisiane et du Cours du Mississippi …* of 1718 covers the greater part of the present-day United States. Both maps are of importance for the Mississippi information, although each add detail to the south-east States in general.

During these first 50 years of the 1700s, the south-east states were being settled – Charleston had been founded in 1672, Mobile in 1712, New Orleans in 1718, Savannah in 1733, for instance – and maps of the area, if not covering a wide area would split, generally, in two ways. Maps entitled *Florida* would extend from Florida along the Gulf coast to Mexico, thereby including Texas and today's southern states; those of *Carolina* would cover the coast and hinterland from today's Georgia to Virginia.

As the century progressed, greater numbers of good maps were issued, many though as separately-issued

Above
British Dominions in America

Thomas Kitchen's detailed map of the
United States.
London c1770
62 x 52cm 24.5 x 20.5in

A detailed decorative map extending, unusually, west and north of the Mississippi and the Lakes, permitting a glimpse of a country waiting to be explored. At this time the southerly States – Virginia, the Carolinas and Georgia had claims to all land west of the Mississippi.

sheets of which few have survived the ravages of time. One atlas publication though which is of particular note is that of Thomas Jeffreys whose *American Atlas*, published by Sayer and Bennett in 1776, included important, large-scale maps of the 13 states of the original Declaration, and Florida, the Mississippi, Canada and South America. Of equally great interest and of greater scarcity are the separately-issued broadsheet maps reporting actions and battles recently fought in the War of Independence (see page 156).

As has been mentioned in respect of state maps of the latter half of the eighteenth century, many periodical magazines included attractive, quite detailed maps which may still be found on the market.

Before leaving the subject of east coast mapping, it would be wrong to ignore the superb charts of Joseph Frederic Wallet des Barres. Des Barres, a British national of Swiss extraction, spent some 18 years before 1774 surveying the north-eastern American coasts, the Saint Lawrence Gulf and Nova Scotia. Returning to England he was instructed to compile an atlas of charts to improve on existing material and, in 1784, he saw published one of the finest chart atlases of any period. With their many engravings of coastal or city profile panoramas, and the use of hachuring and shading techniques to indicate relief, these are magnificent charts. For the *Atlantic Neptune* des Barres utilized where possible his own work, but for the southern coastlines, which he had not himself travelled, he used the most up-to-date reports and surveys from British mariners.

Maps west of the Mississippi before 1800

Development across the continent took centuries after the initial East Coast settlements were established. Saint Augustine in Florida was established in 1565, Jamestown, Virginia in 1607, New Plymouth, Massachusetts in 1620, and later Detroit in 1700 and New Orleans in 1718. European knowledge of the vast North American interior was limited to a few miles either side of river courses and, in the south-west regions, to where Spanish and later French missions had been established.

Atlases of the seventeenth and eighteenth centuries reflected this knowledge, or lack of it, directly in the maps they contained. Accordingly, much early cartography – true or false – of the States west of the Mississippi and north of New Mexico has to be studied in general maps of the Americas or North America. Until the Louisiana Purchase of 1803 and the travels initially of Lewis and Clark, and Pike, serious cartographic detail was limited to the reports from Joliet and Marquette of the 1670s in the upper Mississippi, and from La Salle and d'Iberville of the 1680s along the lower Mississippi and Gulf coasts. Cartography of the

American West is, consequently, dominated by two dramatic misconceptions – California proposed as an island, and the fabled *Mer de L'Ouest* with the corresponding channels from the Pacific north-east towards Hudson Bay.

The first map to concentrate on the south-west area was the 1548 Gastaldi map, reissued in 1561 by Ruscelli. A map of Mexico by Ortelius published in

---

Below
Granata Nova et California

The first map to concentrate on the Baja and any part of California. From Wytfliet's *Descriptionis Ptolemaicae Augmentum.*
Louvain 1597
29 x 23cm 11.5 x 9in

Wytfliet's atlas – the first to concentrate on the New World – combines detail from various contemporary and earlier sources. His sectional maps utilize the work of Mercator, Ortelius and Plancius. This map shows the coast that Sir Francis Drake would have passed in his circumnavigation of eighteen years earlier.

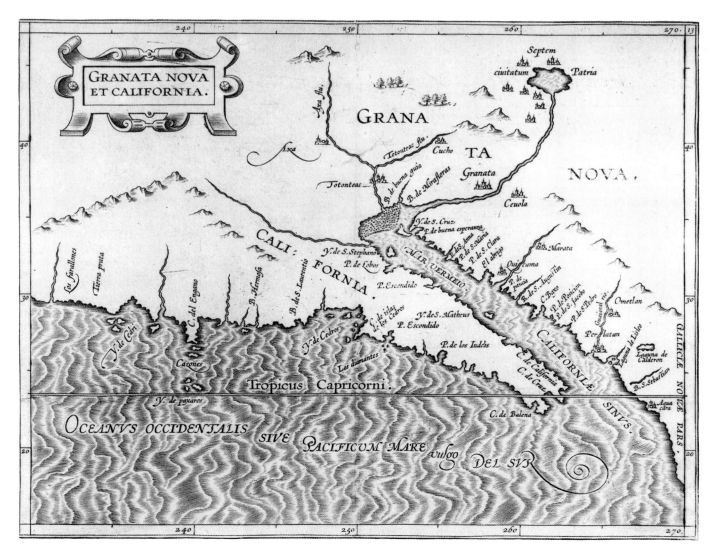

California as an Island

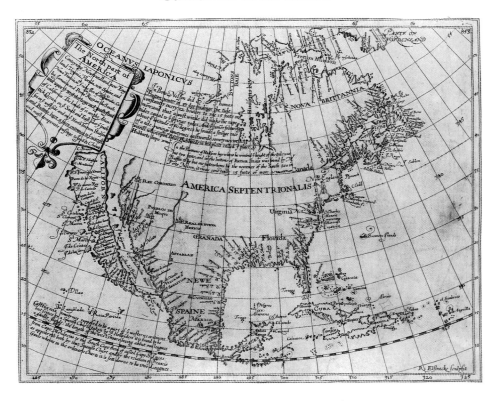

One of the most notable of all carto-graphic errors, and one of the most identifiable, is seen on those maps which show the North American west coast with California as an island. Numerous maps exist which demon-strate this theory which was seriously held by many map makers for approxi-mately one hundred years after 1625.

In 1625 Henry Briggs had produced a map which was included in the compilation of travel reports by Samuel Purchas, published in London. A legend in the lower left corner of Briggs' map refers to a 'Spanish Charte' which identifies California 'to be a goodly Islande …'. Although Briggs' map is interesting as one of the earliest maps of North America alone and for its detail of the north and eastern coast-lines it is for the outline of California that it is best known.

The 'island' theory emanated from Spanish reports identifying the mouth of Baja and a great bay in the north which were presumed to join. The navigator Juan de la Fuca was respon-sible for some of the wildest carto-graphic misconceptions ever mapped. It was his report, in 1592, of the large opening in the American west coast which prompted this theory and numer-ous others, current for nearly 200 years, suggesting the possibility of a channel linking the Pacific to the Hudson Bay

The North Part of America

Henry Briggs' influential and innovative map, showing the island of California.
London 1625
25.5 x 28.5cm 10 x 11in

Engraved by Reynold Ellstracke, one of England's better map engravers of the period, this map was compiled by the mathematician Henry Briggs to accompany his treatise published by Purchas.

and thence to Europe, i.e., the long-sought north-west passage.

Briggs' map became the standard out-line, being used by Speed (see page 136), Jansson (see page 144), Sanson in 1650 (see page 147) and many oth-ers until in 1656 Nicolas Sanson pro-duced a new map showing a different outline (broader and with an indented north coast) which was either copied in its entirety or combined with the original outline on many maps through-out the rest of the century.

Strangely, before the insular theory was promulgated, California had been shown correctly as a peninsula on all maps of the World, the Americas and the Pacific. Even after 1625, many map makers did not accept the theory, or at least displayed an open mind – not by leaving the area blank, but in many

atlases showing one theory on the World map and the alternative on the map of America. Even Joan Blaeu in the *Atlas Maior* demonstrates this – the map of America (first engraved *c*1619) not being altered, whereas the World map shows the island theory.

Sanson's map of 1656 is the first to concentrate on the California/New Mexico region and, during the next decades, maps showing the area in iso-lation are found in sea atlases by Doncker, Van Loon, Goos, and others, and in many land atlases by Sanson, Morden, Mallet, du Val, Coronelli, de Fer and so on. In addition to maps of California the collector of island-theory maps should look at maps of North America, the Americas, the Pacific and the world – all of which may show dif-fering detail or presentation.

Although maps showing the island continued to be engraved and published well into the eighteenth century in Holland, Germany and England, maps contesting the theory were being pub-lished in France from after 1705 when the Jesuit Missionary, Father Kino, by walking from New Mexico to the Cali-fornia Pacific coast, confirmed that California was indeed part of the North American mainland. However, it was not until mid-century, in 1747, that King Ferdinand VII of Spain decreed that California was not an island.

Left
Le Nouveau Mexique ...

Nicolas Sanson's new outline for the island of California became used in place of the flat-topped Briggs model.
Paris 1656
53 x 31cm 21 x 12in

Besides the influential shape for California, this map has importance as one of the earliest to concentrate on these southern states in detail. By condensing the continent south of Lake Erie, Sanson was able to disproportionately enlarge French claims to new territories against the competing Spanish in the south and English in the east.

1579 and based, probably, on Spanish sources, was copied by de Bry, Mercator, Blaeu and many other publishers of the period. Cornelis Wytfliet's atlas of 1597 included a map of Florida and the Gulf Coast (after Chaves' map published by Ortelius) and a map entitled *Granata Nove et California*, the first to concentrate on the Californian region.

Until the mid-seventeenth century and the publication of Nicolas Sanson's *Le Nouveau Méxique et La Floride ...* (see above), there were no new maps of the area alone, apart from Sir Robert Dudley's detailed charts of the Gulf and Pacific coasts issued in his *Arcania del Mare* of 1646. By this time many maps of North America and the Americas had been issued with some detail of the south and western states, but that detail tended to be derivative rather than original (see Briggs' important map of 1625 on page 160 and others on pages 140 and 143 for comparison).

During the 1680s and 1690s the Jesuit missionary, Father Eusebio Kino, travelled extensively around the area of New Mexico and Lower California and in 1701 produced a manuscript map indicating that California was quite definitely not an island. It was mid-century before map makers generally agreed on the peninsula form of California, by which time many fascinating maps and charts had appeared in atlases from all map-making countries. The earliest atlas maps to show Kino's cartography were those by de L'Isle, but it was the French who developed the other great curiosity of west coast mapping in their imaginative plotting of a passage leading from the Pacific across North America and Canada, northeasterly towards Hudson Bay.

French manuscript maps of the 1670s propose a vast flowing river joining the Mississippi to the Pacific. Baron Lahontan in 1703 wrote and produced a map of the 'River Longue' extending westwards from the head of the Mississippi almost, on some versions, to the west coast. Other maps of the period either slant the Pacific coast directly northeasterly, making no allowance for land north of California, or locate an enormous sea within the northwesterly regions, while others depict a series of channels flowing northeasterly. Such fanciful mapping was, of course, one of the most notable features of French 'theoretical' cartography, as it is termed.

Such details continued to appear on maps until accounts of the voyages of Bering and Chirikov were published in 1758, and more significantly the results of Captain James Cook's third and last voyage along the north Pacific coasts of America and Asia between 1776 and 1779 were made known. Between 1792 and 1794 Captain George Vancouver completed Cook's work and confirmed the shape of the North American Pacific coasts.

Within the North American continent, national political and military strife necessitated accurate mapping of the Great Lakes and the Saint Lawrence southwards, but the areas west of the Mississippi lay virtually undisturbed until the nineteenth century when the settlers' wagon trains, and later the railroads, crossed from coast to coast.

Nineteenth-century North American maps

Although lacking much of the decoration and charm of earlier cartography, the maps of North America from around 1800 tell a story of the settling of vast areas of land from the western boundaries of the original eastern states across country to the Pacific coastline.

By the year 1800, the original 13 states, co-signatories to the Declaration of Independence in 1776 had been joined in statehood by Kentucky and Tennessee. In 1803, Ohio state was established and the 'Louisiana Purchase' was made between President Jefferson and the French. Almost a million square miles of, for the great part, virgin territory was purchased for some fifteen million dollars. From New Orleans to

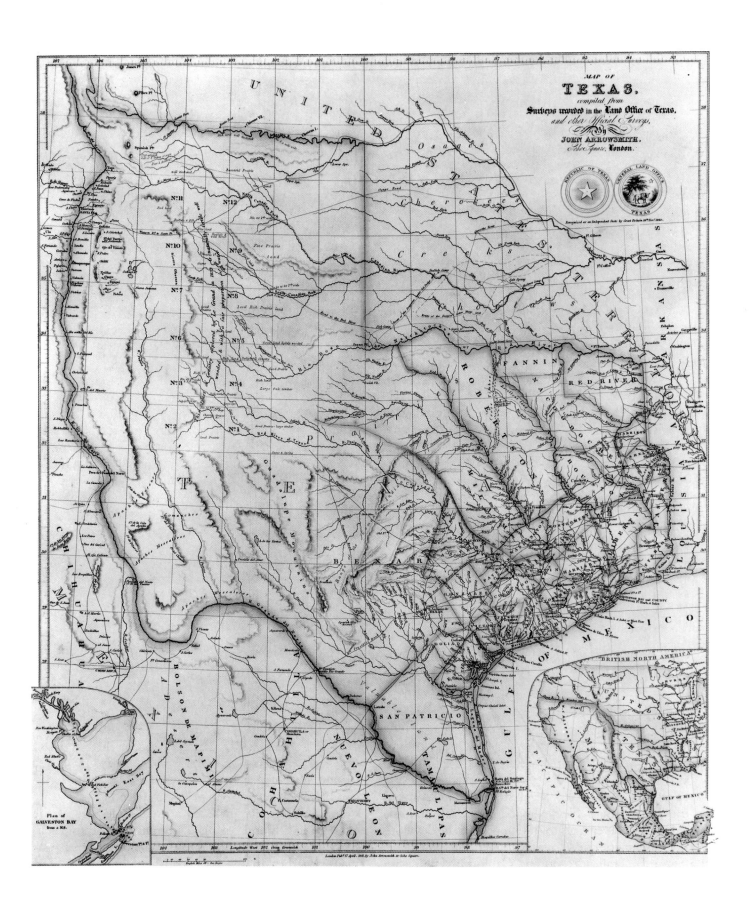

Aaron Arrowsmith's maps of North America and Mexico of around 1800 were amongst the most detailed of the time. After his death in 1823, his business was continued by his nephew, John, who produced in 1841 'The London Atlas'. As well as important maps of Australia and New Zealand, the Atlas included this map of the Republic of Texas, separate from the map of the States. Insets show a plan of Galveston Bay and of the geographical relationship of Texas with Mexico and the United States. The name 'Texas' had first appeared on a map by de L'Isle in 1718 and early maps with this name and subsequent ones of the region alone are now particularly sought after.

Canada and from the Mississippi to Sante Fe, this new land was unknown and required mapping. Accordingly the exploration of Meriwether Lewis and William Clark from 1804 in the northern regions, and Zebulon Pike in the south from 1806, played a significant role in establishing the character and shape of this land.

English map makers were particularly highly regarded at this time and many editions of the large-scale detailed maps of Carey and Arrowsmith especially, traced the expanding knowledge of the interior as reports reached London. At the same time the American map publishing industry was developing fast, but products from this period, important in their topicality, are very difficult to obtain.

From 1795 Matthew Carey's *The American Atlas* was published, the first to be issued in America. Henry Tanner from 1818, Henry Carey and Isaac Lea 1822, Samuel Augustus Mitchell 1846 on, Joseph and George Colton 1855, and many others produced atlases of State maps and other parts of the world. The latter publications, by Mitchell and Colton and slightly later by Johnston, used intricate border frames and sometimes vignette views to decorate the maps.

Many of the above-mentioned map makers produced large-scale, separately issued maps of individual states or the whole United States as it developed. Accordingly, the extremities of such maps are frequently of most interest as the frontiers were pushed back.

By 1850 California, a state as of that year, and Texas, a state of some five years standing, were linked by the Territories of New Mexico and Utah and the subsequent achievements to statehood and the development of the major cross-country routes, the Gila Trail, the Sante Fe Trail, Oregon and California Trails, can clearly be seen on maps.

Besides the US-produced atlases with state maps of all parts of the United States and the Americas, many European-produced atlases included detailed maps of these areas. The atlases by John Thomson (1814) and John Pinkerton (c 1815) include a general map of North America with more detailed ones of Canada, the Northern States, the Southern States, and Spanish North America (i.e., New Mexico, Southern California and the Texas region). The atlas of the 'Society for the Diffusion of Useful Knowledge' (c 1840), includes some detailed maps of sections of Canada and the United States. Another series, which is especially popular for its decorative borders and vignette scenes, is that by John Tallis published in London and New York (c 1850) where North America, Canada (its Maritime Provinces in two maps), the United States, Mexico and California, are treated as separate maps.

The history of exploration and the development of North American cartography, whether of the States or of Canada, is a study in its own right and (given the relatively recent development of the countries), there are a large number of interesting maps available to the collector, not always at great prices.

THE WEST INDIES

The cartographic history of the West Indies reflects, to a degree, the cartographic history of the New World. The first landfall in the western hemisphere of relevance to this study of maps, was by Colombus in the Bahamas in 1492. Accordingly, the first printed map of any part of the New World, by Waldseemüller, published in 1513, shows at its centre point the West Indian islands discovered by Colombus and his successors.

Within 35 years of Waldseemüller's publication Giacomo Gastaldi's atlas, published in Venice, contained maps of parts of the New World including maps of the islands of Cuba and Jamaica together, and Hispaniola/Santo Domingo. From 1561 these maps were reissued on a slightly larger scale. Wytfliet's atlas of 1597 also featured maps of these islands as did Barent Langenes' *Caert-Thresoor* of 1598 whch, later, provided the maps for Petrus Bertius' *Tabularum Geographicarum*. However, besides a small number of other maps of these islands, there were no others of individual islands during the sixteenth century except for Tomasso Porcacchi's *L'Isole più famose del Mondo* published around 1600 which had, in addition to maps of Cuba and Santo Domingo, one of Puerto Rico.

In 1580 Ortelius included a small-scale map of the West Indian islands in his *Theatrum* and, from 1606, Jodocus Hondius incorporated a sheet comprising six maps together in his issue of Mercator's *Atlas*. The islands shown are Cuba, Santo Domingo, Jamaica, Puerto Rico (St Joannis), Margharita and a plan of Havana.

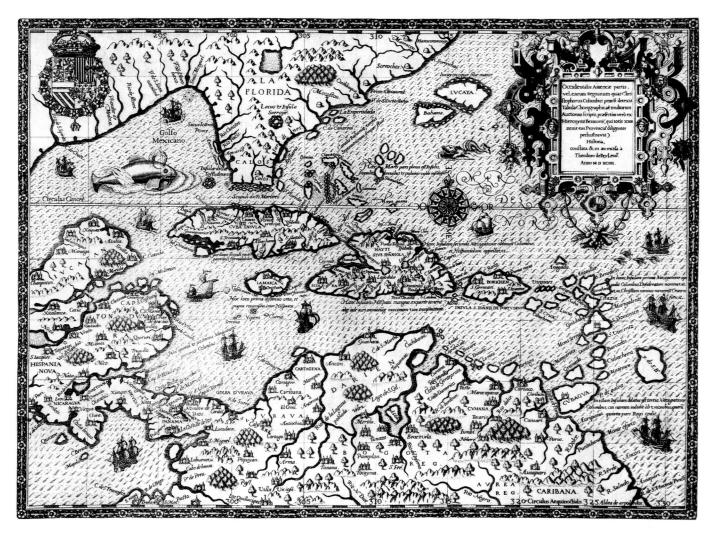

Above
Occidentalis Americae Partis ...

A superb map from Theodore de Bry's
Grands Voyages.
Frankfurt 1594
48 x 33cm 19 x 13in

Published by de Bry, this map accompanied the reports of
Girolamo Benzoni, an Italian who spent about fourteen years
travelling through Spanish possessions in the New World. His
journal was the first non-Spanish view of that Empire. The map
is beautifully designed and engraved and very scarce.

One of the most spectacular maps of any part of the world from this period is the map *Occidentalis Americae Partis* which appeared in Theodore de Bry's *Grand Voyages.* This map, covering the West Indies, Florida, Venezuela, Colombia and greater Central America, is one of very few of the period to concentrate on this area. From about 1625–1630 maps of the area as a whole may be found, but there were still only a few of the islands which are separately mapped.

During the seventeenth century, Dutch sea atlases included charts of groups of islands. The most detailed of these were the work of the Van Keulens, but Jansson, Goos, Doncker, Colom, and even Blaeu also produced maps and charts although they rarely gave any appreciable island detail. During this period many of the islands underwent changes of European ownership so that by about 1700 certain islands were only mapped by specific nationalities. Thus, maps of the islands of Santo Domingo, Martinique and Montserrat appear more frequently in French atlases; Jamaica, Barbados, Tobago and Antigua in English atlases.

As British Imperialism spread throughout the world so the English map-making industry improved, and during the eighteenth century some fine large-scale maps of many of the British possessions in the western world were produced. Amongst these the maps by Mayo of Barbados (1722), Baker of Antigua (1748-9) and Browne of Jamaica (1755) are particularly worthy of note for their detail and fine engraving. By mid-century the atlases of Moll and Bowen, the sea atlas of Mount and Page, and many periodical magazines offered a good range of maps of British-held islands and details of ports and harbours of topical interest within the area.

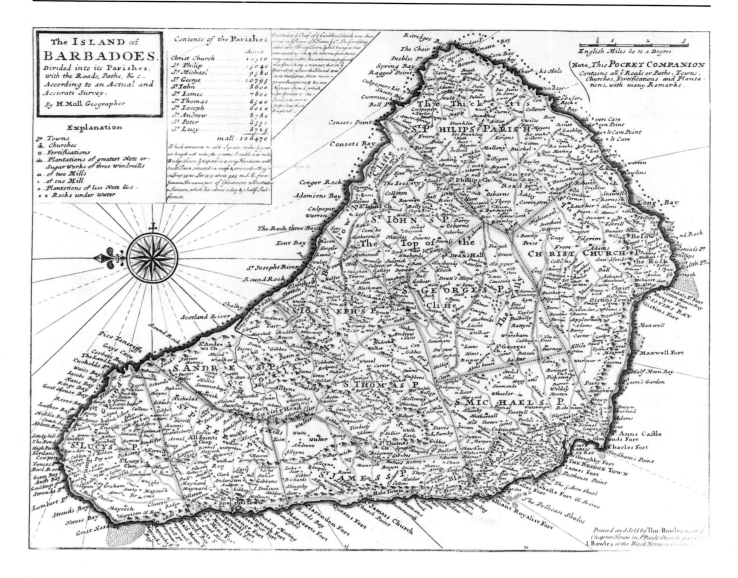

Above
The Island of Barbadoes

A detailed map by Herman Moll.
London c1730
37 x 29cm 14.5 x 11.5in

A detailed English map of one of its prized possessions in the West Indies. Based on William Mayo's major survey, Moll's map was first issued in the 'Atlas Minor' of 1724. The 'Explanation' below the titlepiece indicates amongst the features the size of sugar plantations and the location of submerged rocks around the island.

Atlases by the Frenchmen (de L'Isle, de Fer, Nolin, Robert de Vaugondy, Crepy, Le Rouge, Bellin and Bonne) also contained detailed maps of French possessions in addition, in some instances, to a general map and other island maps of topical interest. Of these atlases the most interesting were those produced by Jacques Nicolas Bellin whose work for *Hydrographie Françoise*, published around 1760, included some large-scale, detailed and decorative maps of all the major islands and whose Volume I of the five volume

Le Petit Atlas Maritime of 1764 includes smaller, but detailed, maps of the islands, their ports and harbours. The complete atlas comprises some 580 maps and charts, of which 56 relate specifically to the West Indian islands.

However, the cartographic progress of the eighteenth century is best seen in the fine maps of Thomas Jeffreys in his *West Indian Atlas*, first published by Sayer and Bennett in 1775. Within this volume there are 39 charts leading from the British Channel across the Atlantic Ocean by way of the Azores, Canaries or Cape Verdes to Bermuda, the West Indies and the encircling Gulf of Mexico coastline. Most islands within the West Indies group can be found here, including the Virgin Islands, a group very rarely found alone on maps, and Bermuda.

By the end of the eighteenth century, the great number of travel books being published invariably included one or more maps of the West Indies and occasionally these maps can be found on the market. Most commonly seen are the plain but detailed maps from Bryan Edward's *History, Civil and Commercial of*

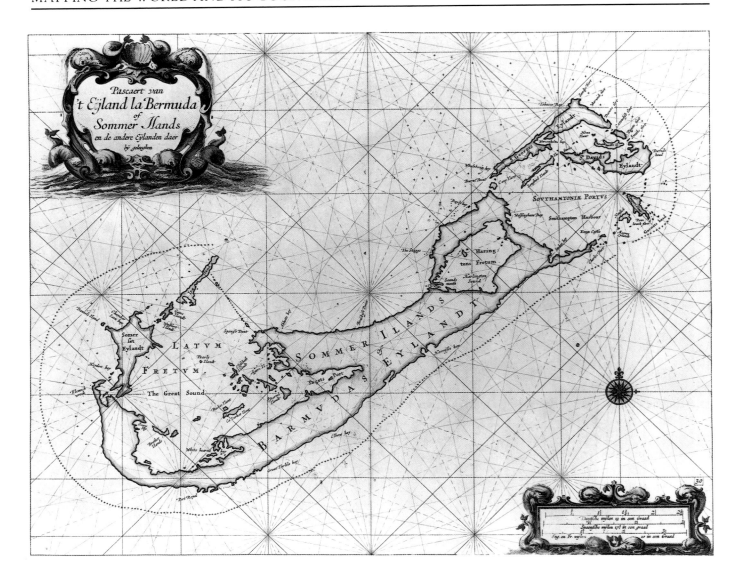

the *British Colonies in the West Indies* of 1794, which included some 10 maps of individual islands. Equally commonly found are the maps from John Thomson's *New General Atlas* (c 1820), which detailed twenty islands on eight sheets. These later maps were reissued with revisions by George Philip (c 1860), by which date the British Admiralty was actively chart-making on a scale previously unseen.

BERMUDA

Although technically not a West Indian Island, Bermuda is frequently seen as an inset detail on general maps of the West Indies, and so is treated as a West Indian Island by most dealers and collectors.

Apparently first visited in 1515 by a Spaniard, Juan de Bermudez, the island's first appearance on an 'available' map is on Gastaldi's 1548 map of the East coast of North America.

The first map of the island alone appeared in Captain John Smith's *The Generall Historie of Virginia, New England and the Summer Isles* first published in London in 1624. This map is rare and has little detail, the

Above
**Pascaert van t Eyland la Bermuda of
Sommer Ilands**

Peter Goos' finely engraved, large-scale
chart of the island.
Amsterdam 1689
52 x 41cm 20.5 x 16in

A particularly popular island among map collectors; there are few maps or charts of the seventeenth century showing detail of Bermuda. The title 'Sommer Islands' is given to the group after the shipwreck in 1609 of Sir George Somers when en route to Virginia. This is one of the most rare charts published of the islands.

engraving consisting of the map in the centre surrounded by a dozen or so views of forts and defensive features set up on the island by the settlers of the Bermuda Company. It was a surveyor, Richard Norwood, who drew this original map and whose work, in 1627, was published by John Speed as *A Mapp of the Sommer Islands ...* in the *Prospect* (see page 46).

Speed's map ran to several editions but is not often found, now, in good condition. Copied by Blaeu, Hondius, Jansson and Ogilby, it formed the standard for the seventeenth century and is cartographically curious for its presentation. The focus of the map is, of course, a large scale depiction of Bermuda, but lightly engraved in the background are the coastlines of Europe and America and, to scale, the island Bermuda thereby showing the island's location within the Atlantic.

Maps of Bermuda may be found issued throughout the centuries and are, particularly in the earlier instances, very decorative and accordingly are very sought after, Maps by Speed, Blaeu, Hondius, Jansson, Moll, Bowen, Bellin, Zatta and Tallis may be encountered but the more sought-after items by Ogilby, Robyn, Van Keulen, Van der Aa, Lemprière, and many others are rarely seen on the market.

MAPPING OF SOUTH AMERICA

The cartography of South America is more clearly defined than that of North America. In broad terms, its Atlantic coasts and southern limit were known after Magellan's voyage of 1519 and its west coast was generally correct by the end of the sixteenth century –contrast this with North America where the northern limits were not known until the late nineteenth century and its west and north-west coast were unknown until the late seventeen hundreds.

The Spanish and Portuguese interest in Southern and Central America and the subsequent expeditions of Cortez in Mexico and the Pizarros in Peru led to early, relatively accurate mapping. Gastaldi's 1548 map, *Nueva Hispania* (the southern areas of North America including today's Mexico and Central America) and his *Tierra Nova* (South America) reflect this interest. Ortelius' map *La Peruviae ...* (see page 143) is one of the few sixteenth-century sectional maps of a part of South America, and it is usually necessary to look at maps of the whole western hemisphere to examine the cartography of this great landmass.

The 1590s saw the publication of two particularly attractive maps of the South American continent. From Theodore de Bry's *Grands Voyages* comes a finely engraved map which makes up in style and design for what it lacks in accuracy (see page 168) and four years later, in 1596, Linshchoten's *Itinerario* includes an equally awry outline, even more flamboyantly decorated with ships, animals, native scenes (including cannibalism) and so on. Both these maps are difficult to obtain.

During the seventeenth century, most map makers featured a map of North and South America with various new maps of specific areas. However, in 1606, Jodocus Hondius included a map of South America *America Meridionalis* with an inset detail of Cusco, and a map of *Fretum Magellanicum* (The Magellan Straits). By 1630 Henricus Hondius had sold to Blaeu further maps of Venezuela, Granada, Chile, Paraguay, Peru and Guayana and these appeared in subsequent Blaeu

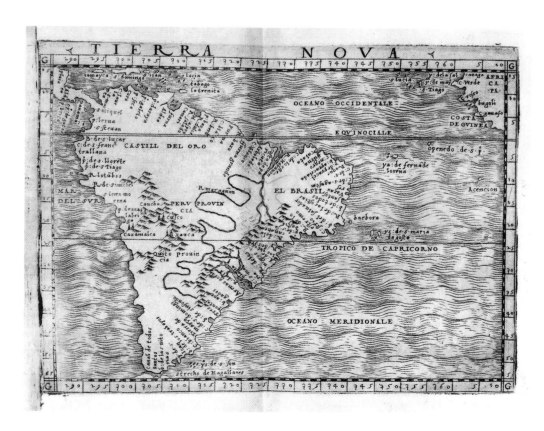

Left
Tierra Nova

Giacomo Gastaldi's 1548 copperplate map – the first to show South America independently.
Venice 1548
17 x 13cm 7 x 5in

Published in Gastaldi's miniature version of Ptolemy's 'Geographia', this is one of the most important maps within the atlas. Versions of the 1561 and a later issue of this map by Ruscelli may be found and are interesting additions to a collection as the earliest, reasonably-priced, obtainable maps of the continent.

atlases. For publication in the *Atlas Maior* of 1662 on, Blaeu had prepared some new, replacement maps and had added a sequence of maps of the Brazilian coastline. Except for the Brazilian maps, these maps published by Blaeu formed the basis of Arnold Montanus's/John Ogilby's maps in their volume on America.

From the outset cartographic detail of the region was limited generally to the coastlines, the greater part of inland South America appearing as impenetrable jungle except those areas – as in Columbia and Peru – where a civilization already existed. During the seventeenth century most cartographic development was in improved coastal outlines, many charts being published in Amsterdam at this time. Even where better information was known to the Spanish, little of this was published due to the secrecy surrounding their colonial activity.

Despite minor improvements, little cartographic progress was made until the journeys of Alexander von Humboldt, around 1800, along stretches of the Orinoco and the Amazon. However, during the second half of the eighteenth century, quite a large number of maps were issued showing separate South American countries and maps by makers such as De Fer, Bellin, Vaugondy or Bonne make interesting collectors' pieces.

Americae Pars Magis Cognita

Theodore de Bry's fine map of the West Indies, Central and South America.
Frankfurt 1592
39 x 36cm 15.5 x 14in

A finely engraved map, illustrating de Bry's issue of Von Staden and Lery's voyages to Brazil in the 1550s. The map is most curious in its shape and bears little resemblance to any other maps of the period.

The map appears in a second edition when it was re-engraved with similar content but altering the date at bottom left to MDCCXIV.

GLOBES, CURIOSITIES AND MINIATURES

CELESTIAL MAPS

The study of the skies, as distinct from our terrestial geography, is a science in itself, and one requiring quite a considerable amount of knowledge to fully, or even half-fully understand. However, despite this specialization, maps, plans and diagrams of the constellations, the planetary systems and other celestial phenomena should be considered within the scope of this work. Many atlas makers included a map of the skies within the volume and these maps are occasionally available to the collector. In certain instances, they are amongst the most decorative charts which may be found of any period.

Acknowledging that such specialized knowledge is required for in-depth study of this field, I make no apologies for the limitation of the information within this writer's compass.

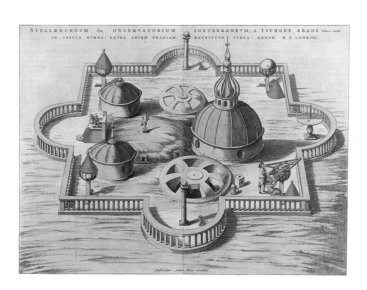

Stellaeburgum sive Observatorium Subterraneum

Tycho Brahe's observatory at Uraniburg
from Blaeu's *Atlas Maior*.
Amsterdam 1658
54.5 x 43cm 21.5 x 17in

Blaeu studied at Uraniburg for some six months in the early 1590s during which time his natural aptitude for cartographic sciences developed. Soon he was producing globes, maps and chart books which formed the basis of his great publishing house. As a tribute to both his father and Tycho Brahe, Joan Blaeu (1596-1673) published in 1658, a set of fourteen maps, plans, illustrations of the equipment, and views relating to the island and the school.

Claudius Ptolemy, whose *Geographia* catalyzed medieval atlas production, can also be regarded as one of the most influential figures in the history of astronomy. His *Almagest*, compiled by 150 AD, catalogued over one thousand stars, defined in 48 constellations. This list was amended or redrawn by other astronomers – the Alfonsine Tables, Copernicus and others – but was not revised until the late sixteenth century when Tycho Brahe at his observatory, 'Uraniburg', recatalogued the system.

Observations of the heavens had been made from the earliest times, and with the advent of printing, books about astronomy, star catalogues and illustrations of natural phenomena (comets, eclipses and so on) were produced. Many of the woodblocks of the *Nuremberg Chronicle* of 1493 deal with such subjects. The earliest printed star charts were designed and engraved by Albrecht Dürer (who also produced an equally rare World map) and were published in 1515. Some seventeen years later Johannes Honter engraved a pair of woodblocks based on Dürer's, but adopted for the first time a view of the constellations as seen from the Earth (see page opposite).

The early part of the seventeenth century saw two publications of particular interest. Johann Bayer's *Uranometria...* of 1603 is generally regarded as the first star atlas. With its revised Ptolemaic star catalogues of Tycho Brahe, the incorporation of a new lettering system for the most important stars (using Greek and Roman letters), and fine copperplates for each constellation the book was particularly popular, being revised some eight times prior to 1689.

Julius Schiller's atlas of 1627 provides an interesting diversion from the mainstream of astronomical symbolism. Technically, Schiller's atlas is an updated version of Bayer's and, as such, was the best available at the time. However, in converting the traditionally accepted Greco-Roman constellation figures to Judeo-Christian characters, the inherent quality of the work could not offset the lack of enthusiasm for the redrawn constellations. Accordingly, only one edition of this work is recorded.

During the seventeenth century some magnificent globes, both terrestial and celestial, were made. Invariably appearing in pairs and in various sizes from a few centimetres across to as large as 60 or 70, superb globes with printed gores were issued by Willem Janszoon Blaeu from the early years of the century and, in the closing years, by Vincenzo Coronelli in Venice.

The highpoint of celestial atlas production, however, and the volume that ranks with Blaeu's *Atlas Maior* and Goos's *Zee Atlas* is *Harmonia Macrocosmica ....* by Andreas Cellarius. Published by Jan Jansson in Amsterdam in 1660, the atlas comprises some 29 star charts and diagrams which portray varying celestial and planetary systems, orbits and theories. Little is known of Cellarius besides the information provided within the atlas, that he was Rector of the Latin School at Hoorn, 20 miles north of Amsterdam. The format of most engravings is similar – a sphere occupying the sheet top to bottom within which the diagram or chart is positioned, allowing up and down each side, decoration of an instructional, symbolic or purely aesthetic nature.

The Cellarius charts, issued in 1660, 1661, 1666 and 1708, occasionally appear on the market and can be found in superb, bright, original colour, highlighted with gold, making them highly decorative items. The later editions of 1708 has the imprint of the publishers Valk and Schenk on each engraving and is typically in relatively subdued, though also attractive, colouring. Many terrestrial and maritime atlases from the second half of the seventeenth century included a celestial chart, usually in two spheres, north and south, and many of these are particularly attractive. Besides double hemisphere maps published by, amongst others, Van

Imagines Constellationum Borealium

Johannes Honter's important woodblock map of the northern skies.
Basle 1532
28 x 30.5cm 11 x 12in

Issued with a matching chart of the southern skies in Henrici Petri's edition of Claudius Ptolemy's 'Omnia, quae extant opera, Geographia excerpta', this is the first map to show the star positions as seen from earth (i.e, inside as opposed to outside the stellar universe).

The standard figuration of the constellation is maintained although earlier Arab charts would show figures in Arab costume, the earlier maps of Dürer show classical figures with few garments anyway, but Honter up-dates the dress of his celestial characters.

Around the circumference of this sphere are the figures we now associate with the zodiacal signs and many other recognizable forms.

The date 1532 at lower left refers to the date of engraving of this block which was issued a number of times later.

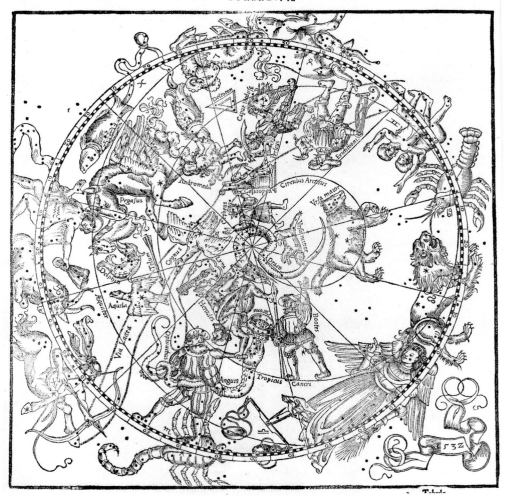

IMAGINES CONSTELLATIONVM
BOREALIVM.

Keulen, Schenk and Mortier in Amsterdam; Rossi, Coronelli (Venice) and, later, Zatta and Cassini in Rome; and Homann, Seutter and Lotter in Germany, there are some particularly decorative, fine, single, 'planisphere' maps, after Van Luchtenberg, published between 1664 and about 1710 by Doncker, Robyn and Danckerts in Amsterdam.

John Flamsteed's *Atlas Coelestis*, first published in 1729, was the result of some 40 years' observations by Flamsteed, England's first Astronomer Royal. The atlas ran to two further editions published in London, and others in Paris and Berlin, and was also issued in quarto, as opposed to folio, format. The other major celestial atlas of the eighteenth century was that by Johann

Right
Atlas Coelestis seu Harmonis Macrocosmica

The superbly decorative title page from Andreas Cellarius' beautiful celestial atlas.
Amsterdam 1660
27 x 44cm 10.5 x 17.5in

This elegant title page represents fully the contents of the book. Each of the charts is well-designed, well-engraved and often, as is the case here, is in fine original colour heightened with gold.

The Muse of Astronomy, Urania, is surrounded by scientists, mathematicians and astronomers and celestial globes and observation equipment. At lower right a bound volume is typical of the fine red morocco binding with gold embossing, used in Amsterdam at the time. Many of the great atlases can be seen bound in this fashion now. Two cherubs hold aloft the book's title on a banner whilst another couple, using crossstaffs, study the zodiacal signs of Libra and Virgo.

Above

Sceno Graphia Systematis Copernicani

One of the least complex diagrams from
Andreas Cellarius' atlas, showing the
relationship of the Earth to the Sun at
the times of equinox and solstice.
Amsterdam 1660-1708
51 x 43cm 20 x 17in

*Peter Schenk and Gerard Valk reissued the copperplates of the
original Cellarius atlas in 1708. These plates, identified by the
addition of the new publisher's imprint on the title banner, are
usually found in the wash colour seen here – pastel shades which
lack the vibrance of the original colouring but which are,
nevertheless, most attractive.*

*Cherubs disport themselves and female figures represent dark
and light.*

Gabriel Doppelmayr, published by Homann in 1742.
More ornate than the Flamsteed work there was,
however, only the one edition, though some of the
maps from this atlas had been previously published by
Homann, bound into atlases or as separate sheets.

Throughout the nineteenth century most of the
larger general atlases included reference to planetary
and celestial charts and diagrams. However, an
increasing number of star atlases, in keeping with the
scientific progress of the day, were being published to
satisfy the demand from a more curious but also more
knowledgeable public.

One particularly attractive series of celestial maps
published to satisfy this demand was that entitled
Urania's Mirror or, a View of the Heavens published
around 1830. This contained 32 cards with engraved
constellations, with each major star identified by a pin
prick. Thus, if the card is held up against the light, the
pin holes stand out as would stars in the night sky.

Above
Globo Celeste ...

One of four sheets of celestial globe gores
by Giovanni Cassini.
Rome 1792
34 x 49.5cms 13.5 x 19.5in

*These large gores would have been pasted onto a globe of some
14 inches in diameter and would be accompanied by a matching
terrestrial globe. Here the classical constellation figures are,
typically, without dress. The imprint in the title cartouche claims
topicality for the information shown being based on the
observations of Flamsteed and de la Caille.*

THEMATIC MAPS

As will already be apparent, the majority of collectors will build their collections on a thematic principle, i.e., concentrating on a particular subject or theme. In most instances this concentration will be on a *specific area, mapmaker* or *period of interest* to the individual. However, there are numerous other categories which, in themselves, are each worthy of special attention.

I summarize a number of these other categories below to show the variety of subject headings worthy of study, many of which receive mention in other parts of this book, and all of which would be ideal subjects for separate books themselves.

Area, age and map maker

By its very nature any map described or illustrated within this book could fall within any of these categories. However, there is complete freedom for the collector to determine the possible limitations to his collection.

Place of publication

A preference or interest may develop around certain groups of printed maps. Thus, examples of Dutch, French or Italian maps may be collected. However, the greater limitation of a time period may be added, for instance, 'sixteenth-century, Italian-produced maps', or 'nineteenth-century English-produced maps'.

Language in which a map is printed

Maps are commonly found in English, French, Dutch, Latin, German or Italian. A collection could, however, be based on maps in languages such as Spanish, Danish, Russian, Hebrew or Armenian which are sometimes encountered. These may have been produced either in their native country or as foreign language editions of more commonly encountered maps. Ortelius's maps appear in Latin, French, Dutch, German, Spanish, Italian and English editions and some of Jacques Nicolas Bellin's maps of the mid-eigthteenth century are found with the original French text translated into Danish, Italian or Dutch.

Map purpose

The majority of maps which will be considered by collectors are topographical in format, i.e., they show the locations of places on land in relation to each other. However, several other forms of maps can be identified to satisfy other specialized requirements.

Sea charts, celestial charts and town plans are each discussed in other parts of this book.

Road maps are best known in the Peutinger tables and in the maps by John Ogilby (see page 84) and his successors. However, throughout cartographic history, there has been surprisingly little emphasis placed on a map style which we take for granted today.

The earliest printed road map known today was by Ehrhard Etzlaub, about 1495, of European routes to Rome – the *Rom-weg*. From the seventeenth century one can hope to find maps of postal routes throughout the continent and the roads of England and Wales, although roads were not shown on general series of county maps until after the publication of Ogilby's *Britannia* in 1675. The playing card maps (1670) and the Camden's Britannia maps (1695) of Robert Morden were the first to show roads in any detail.

Eighteenth-century road maps of England and Wales and, particularly, France can be found. In the form of strip maps, these are often the largest-scale maps of localized areas available for the time so should not be overlooked by collectors.

Canal, waterway and railway maps

These products of the late eighteenth and early to mid-nineteenth century are records of the birth, in England and Europe, of the Industrial Revolution and, in North America, of the opening and developing of a continent. Both subjects provide a fascinating study of social history though the maps are not particularly decorative. England's first canal was built by James Brindley from 1761 between Worsley and Manchester and the first railway line to appear on a map was the Surrey Iron Railway between Wandsworth and Croydon, built in 1803.

Railway maps of many parts of the world may be found in their original state – folded into covers, probably with timetables and a small guidebook – and were produced in great numbers. However, relatively few have survived in good condition owing to natural wear and tear.

Physical and geological maps

As with the earlier categories these more scientifically-based maps become evident during the nineteenth century when a number of 'Physical' atlases were produced – most notably by W and A K Johnston. The maps within such atlases might define ethnographic, meteorological, zoological or botanical distribution information. Ocean currents, prevailing wind patterns, volcanic and seismological information might be plotted, and so on. Such maps reflected a growing public interest in these rapidly developing sciences although

some such subjects had been touched on during the earlier centuries, notably by Athanius Kircher, who, in his *Mundus Subterraneus* of 1665, produced maps illustrating theories of ocean currents, volcanic distribution and the core structure of the Sun and the Earth.

Philippe Buache, from around 1750, produced several geological maps identifying surface deposits of parts of France. The technique became refined by the early 1800s and the publication of William Smith's map of the strata of England and Wales, in 1815, further established the colour-band delineations which we recognize today.

Pocket maps and wall maps

Each category, at opposite ends of the spectrum size-wise, and designed for specifically different purposes, is dealt with elsewhere (see pages 180 and 181).

Battle plans – military and naval

These were mainly published in history books or as separate, broadsheet publications. In the latter instance they would often be the equivalent of a modern newspaper report and, as such, the most up-to-date information available to the public. Of particular

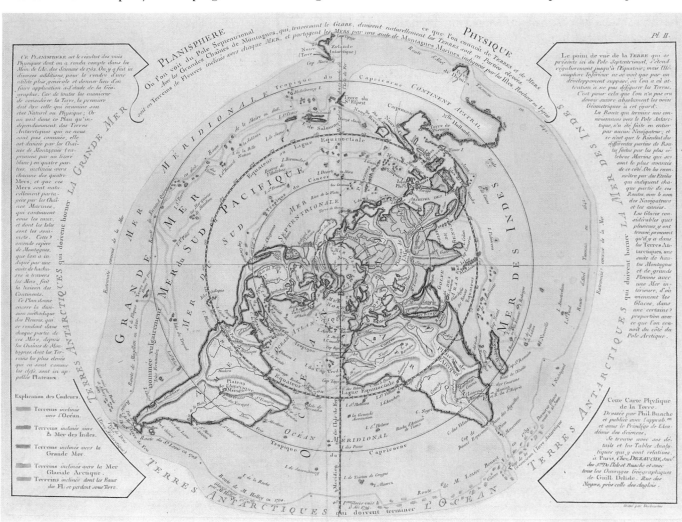

interest are the plans of battles issued by Faden, Sayer and Bennett in London of the American War of Independence.

Large-scale or miniature maps

Separately-issued, large-scale maps of the counties, or individual countries can, of course, provide far greater detail than many of their atlas counterparts. Ultimately, the Ordnance Survey in Britain and national surveys in many other countries were to satisfy this demand, but there are some particularly fine, large-scale national maps which could be included, for decoration and style, with wall maps.

Collectors of 'miniature' maps have their own size definitions of 'miniature'. Originally, the distinction would have been that maps from pocket atlases were regarded as 'miniature'. In this category the world maps by Gastaldi (1548), Ortelius (from 1577), Langenes/Bertius (from 1598), and many others of later years would be included, none more than postcard size. Equally, the English county maps of Pieter van den Keere (from 1619), John Bill (1626), and many others would qualify.

In more recent years the definition has expanded to include those maps from 'reductions' of larger atlases. Thus, the *Atlas Minor* maps of Mercator, Hondius and Jansson qualify, although these might be large as about 25 x 20 cm 10 x 8 in.

For many collectors the appeal of the smaller size map, as opposed to full folio, is simply one of convenience. A large quantity of small maps can be stored in albums or on bookshelves, as opposed to portfolio cases or plan chests. Accordingly, one finds that many stamp collectors (whose interests and aims in collecting are not dissimilar to those of map collectors) will build a collection of miniature maps to complement, and keep with, their collection of stamps.

Map curiosities

The range within such a category is extensive and discussed more fully from page 186.

Decorative

Regardless of age, author or subject matter, maps may be collected simply for their decorative content. More specifically, a collection could be based on maps showing monsters (e.g., Ortelius's map of Iceland, see page 14), native characters, particularly elaborate title or decorative cartouches, finely detailed ships or town depictions, etc. The variety is limited only by the reader's, or observer's, imagination.

'Bird's-eye' format

More typically seen as a style of presenting a town plan, there are map engravings which adopt this combination of overall survey with a topographical view. The best-known example is probably the Bickham series of county maps although there are many other examples to be found.

Above
Typus Orbis Terrarum

A miniature of Abraham Ortelius's oval
World map.
Antwerp *c*1600
11 x 8cm 4.5 x 3.25in

A charming miniature map in original colour based on the later plate of Ortelius's World map in which the most notable alteration was the correction of the South American coastline.

Numerous editions of this pocket-atlas of Ortelius's maps were published in Antwerp and, later in the seventeenth century, in Rome.

Maps from other 'miniature' versions of folio atlases might be found based on Speed and Hondius/Mercator; and from small atlases by Gastaldi, Magini and Porcacchi, published in Italy; and by Morden and Seller in London. From the eighteenth century there is a wealth of material for the collector of provincial and national maps produced in France, England and Germany.

Left
Planisphere Physique

One of Philippe Buache's numerous
physical maps.
Paris *c*1760
44 x 34cm 17.5 x 13.5in

This particular map, with its complex North American outline and greatly extended Australia, attempts, by identifying the major mountain chains of the world, to show the catchment areas of each of the oceans. Each chain acts as a watershed and directs the flow of the major rivers.

MAPS AS OBJECTS

The emphasis of this book, as I hope will be apparent, is on those maps which are available to collectors and decorators and which are readily found on the market in specialist map shops and galleries. By far the greater proportion of these were originally issued printed in atlas or bound form. However, there are other instances where maps, printed on paper, were produced for specifically different purposes. The following chapter attempts to identify some of the 'alternative' forms of printed maps and also suggests a few more unusual maps which have not been printed and are sometimes available.

Cartographical misconceptions

As distinct from cartographical curiosities, this category includes the more extraordinary cartographical representations which were, when promulgated, seriously accepted as fact. The most obvious example is the theory and illustration of California as an island (see page 160). Some of the French theoretical cartography and the subsequent outlines of a 'north-east' passage (see page 48) are also particularly strange.

Pocket maps

One of the great inconveniences of the standard folio size atlas was, obviously, its lack of portability. Over the years this problem was resolved in two ways. The development of the so-called 'pocket' atlas saw either the traveller's practical guide which could be packed conveniently, readily to hand, for reference when 'en route', or a reduced format version of a previously larger, more expensive atlas, intended to satisfy demand from the 'popular' market. In the first instance the road books of the eighteenth century are obvious examples; in the second, i.e., the small format versions of previous publications, examples include Ptolemy's *Geographia* by Gastaldi in 1548, Humble's edition of Van den Keere's county maps from 1627, or Blome's *Speed's Maps Epitomiz'd* of 1681. Maps from each of these types of atlas are often offered to collectors.

Pocket book maps

The second, and more interesting, development along this line is that of the 'pocket-book' map. This usually took the form of a single map, often dissected, mounted on linen or extra thick paper, and bound into a

protective folder or loose in a slip-case. It is not uncommon to find maps taken from atlases, then mounted and folded, for practical use during the seventeenth century. Speed's maps of the counties are sometimes found in this state, conceivably having been intended for use 'in the field' during the Civil War *c* 1642–43. However, the more interesting maps of this genre were those specifically separately issued. These maps, frequently on much larger scales than those issued in atlases, were intended for practical use and, accordingly, are not always in the best condition.

Published extensively throughout the nineteenth century, these large-scale maps had their origin in the previous century which saw the production of more detailed and more accurate maps than ever before, whether of English counties or of foreign parts. One of the most interesting factors regarding this type of map is that, since they were intended for practical use, they were revised more frequently than their atlas counterparts, and so developments in communications, for instance, or the latest discoveries in distant countries can be noted.

The nineteenth-century interest in travel – the birth of commercial tourism which we take for granted nowadays – spawned a dramatic increase in town plans and guides of major cities or resort areas, not just in Britain but throughout each continent of the Old World. Pocket maps and guides can be found relating to the New World, either for the use of those who already lived there, or as guides for the great numbers of emigrants who sailed west to North America.

Folding maps have, to an extent, been overlooked by collectors in the past since they do not conveniently 'fit in' to a folder or drawer of flat maps, and are not

usually of sufficient decorative quality to be framed and hung. Accordingly, there is a relatively wide range of interesting maps available, and although the rarities fetch high prices, many can be found at a reasonable cost.

Wall maps

At the opposing end of the spectrum sizewise are maps prepared specifically to be hung on walls. Invariably large, probably made up of several paper sheets pasted together on a backing board or cloth, these maps frequently combine detail and scale with spectacular decoration and were designed particularly with the latter feature in mind.

Unfortunately, the large size of such items almost inevitably leads to damage and very few examples of this type of map are ever found in perfect condition. However, given this fact, it is still possible to find

Above
L'Amerique

A fine French wall map of the eighteenth century.
Paris 1740
137 x 121cm 54 x 47.5in

The grand scale of this map and its elaborately designed border indicates its original purpose as a wall map. This particular map would have been issued with others of each of the continents and the world. Each of these maps would be surrounded by scenes, thirty in this case, depicting various stages in the history of that continent or notable physical features or elements of native life. The large title cartouche includes a dedication to Louis XIV and the publisher's imprint indicates that this is a relatively late edition of the map.

Rarely found in good condition, most maps of this type must not be examined too critically – even with faults they are still often spectacularly decorative items.

reasonable copies of these rare maps. Most of the great Dutch and French map makers produced wall maps, but, it is French maps by de L'Isle, de Fer, Nolin, Clouet and others, that are the most commonly found.

Such maps would normally hang with wooden bars or rollers at top and bottom as opposed to being framed, and in fact some Dutch artists of the seventeenth century used wall maps as back drops to their portraiture work, Vermeer probably being the best-known exponent of this practice.

During the latter half of the eighteenth century, improvements in mapping techniques and new demands on account of military, political and commercial interests led to a dramatic increase in the production of maps designed for wall use. The London map trade was particularly active in this respect and the maps of John and Aaron Arrowsmith, which were used well into the next century, are probably the finest examples of the period.

Globes

Wall maps, produced very much with their decorative nature in mind, were treated as art objects as much as maps, and this is equally true of globes.

Produced in quantity from around 1600 onwards, terrestrial and celestial globes can be found in many sizes and styles.

Miniature globes can be found as small as 1 inch in diameter, but most commonly encountered are the pocket-globes approximately 3 inches across. They were kept in an animal, fish or reptile skin case within which a chart of the heavens is pasted. In vogue during the eighteenth and early nineteenth century, these charming items are now greatly sought after.

Larger globes were intended for desk- or table-top

Right
Le Globe Artificiel et Mecanique ...

A fascinating and most unusual
educational globe.
Paris c1820

Folding flat within a four-leaf wallet, with one loose descriptive sheet, this curious six-gore segmented globe operates on a 'draw-string' principle.

Owing to the fragile nature of such items, they are rarely found in good condition and complete. The folder contains geographical notes concerning the world and the solar system. When opened, the globe measures about 15 cms (6 inches) across and would have been intended for use in schools and colleges.

Left
An Accurate Map of 460 Miles Round London

A revolving mileage chart
engraved by Emanuel
Bowen and sold by
Benjamin Martin.
London c1740
28.5cm 11.25in diameter

A fascinating and particularly uncommon item. The map is pasted onto board and revolves within a brass scale indicating distances from London. On the reverse are a series of diagrams indicating the Earth's orbit around the Sun and the relationship to Saturn and Jupiter.

Bowen was one of the most prolific mapmakers of the period and Martin one of the foremost scientific instrument makers. As can be seen, understandably the map has suffered some wear but is, nonetheless, a most interesting cartographic object.

Below
The Balance of Trade Plate

A decorative commemorative
cartographic object.
English 1887
24cm 9.5in diameter

An amusing and fascinating decorative plate produced to commemorate Queen Victoria's fiftieth year on the throne in 1887. The coats of arms of Australia, Canada, India and Cape Colony are shown, as are vignette views of 'Australians in the Soudan' and 'Canadian Voyageurs on the Nile'. Britannia is seen surrounded by respectful inhabitants of the Empire whose number – it is stated – is 305,347,924.

use. In the case of smaller examples, they would normally be on a single column support (usually wood, sometimes metal); those over 8 or 10 inches in diameter were set into a four-legged cradle support.

The largest globes, from about 14 or 16 inches were floor-standing and invariably had finely designed, turned and crafted 'furniture' – the term given to the stand or support.

The circular globe itself consisted of a wooden framework skeleton with a skin of 'papier-mâché' type pulp, or cloth and plaster built up around the frame. Paper 'gores', on which the map detail was printed, were pasted onto the globe, coloured and, in more recent times, covered with a 'lacquer' or varnish coating for protection.

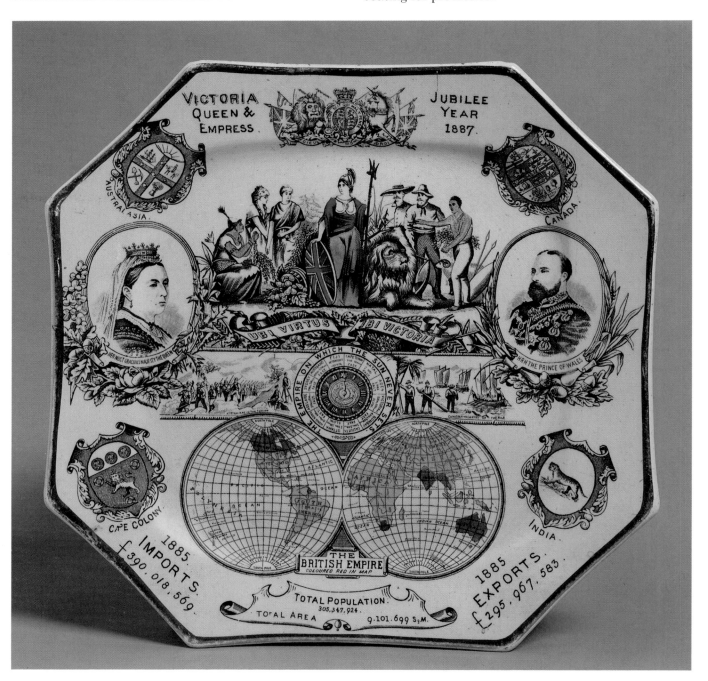

Unfortunately, few globes which come onto the market now are in perfect condition. Staining or discolouration, wear and damage are often encountered and a knowledge of antique furniture is necessary to ensure that all parts of the cradle or support are original or without repair.

As with wall maps many of the major map makers of their periods produced globes. Leading figures were the Blaeus in Amsterdam, Coronelli in Venice, and Adams and Cary in London. As each country enjoyed periods of intense map and atlas production, so a comparable industry making globes appeared.

Cartographic and geographic accessories

As we shall see in the chapter on cartographic curiosities, there are further areas for the collector to explore beyond the range of atlas-maps. Here, however, I mention some other cartographic objects which were produced as serious instruments.

Maps for learning

Inevitably during the last century, maps became an essential part of a child's education. As such wall maps, globes and atlases were produced for use in schools, but given everyday school use they are rarely found in good condition.

In order to make geography a more amusing subject, jigsaw puzzles (occasionally with pieces the shapes of counties or countries) were produced. Instructional card games incorporating maps and information are also found. Unfortunately, puzzles and games are rarely found complete but, nevertheless, they provide a charming insight into the upbringing of, for instance, the Victorian child. Futher diversions along the educational map route are seen in the next chapter.

Map samplers

Yet another decorative map object which might be seen – in general antique shops rather than specialist map or book shops – is the map sampler. Dating mainly from the mid-eighteenth to late nineteenth centuries these maps, often finely sewn on silk or cotton, show the world or the seamstress's country and originally served two purposes – firstly, to teach the young lady how to sew and secondly, to act as a geography lesson.

These elegant by-products of a former age are often charmingly decorated within floral-design frames.

Commemorative maps

Of interest to some collectors – more as ornaments than for their cartographic content – are pieces of china-ware, such as plates and mugs on which maps are incorporated into the design. Most commonly found from the Victorian era, these invariably commemorate particular anniversaries, for example, Queen Victoria's fiftieth year on the throne or the Relief of Ladysmith.

Silk or cotton scarves can also sometimes be found produced for the same reasons.

Collectors of such 'memorabilia' can also find charming 'souvenir'-type items incorporating maps within the design, or even map-shaped. Amongst these oddities are such items as ashtrays, crockery, inkstands, letter openers, and so on. Most such items originate from the last century but were still produced in great quantities earlier this century. Accordingly, they are not necessarily antique but, nevertheless, are amusing and decorative cartographic objects.

185

CURIOSITIES

As distinct from cartographical objects designed as ornaments or to instruct, and as well as the strictly topographical atlas maps, there is yet a further category of maps. The cartographical curiosities identified below exclude wild cartographic misconceptions (these were believed at the time and were therefore serious maps), and also serious cartographic objects previously mentioned. The curiosities listed below were designed for the maker's amusement, for the reader's amusement or for the pupil's amusement.

The earliest curiosities that the collector is liable to encounter are the famous maps from Heinrich Bunting's *Itinerarium Sacra Scripturae*, published in 1581. In addition to correct maps of the Holy Land are maps of the World in clover leaf form with Jerusalem at the centre, Europe as a robed female figure, and Asia as Pegasus, the winged horse (see below). The concept of the European figure-map had, in fact, already appeared in 1537, but the Bunting version and another in Munster's book (*c* 1580) are more commonly found than the original by Joannes Bucius.

Omitting other particularly rare maps of the World, the next and arguably the most famous curiosity is the sequence of *Leo Belgicus* maps issued in Cologne from

1583 and then the Low Countries in a variety of forms until the eighteenth century (see opposite).

During the seventeenth century the development of the curiosity (within our definition) was relatively limited compared with the proliferation of publications which followed in the eighteenth. However, two particular cartographic types are worth noting. In 1590 and around 1605, William Bowes had produced sets of playing cards which included, within their design, maps of each county of England and Wales. However, copies of these cards are known in only a handful of examples. From 1676 Robert Morden and William Redmayne published sets of cards with county maps thereon and these do appear on the market occasion-

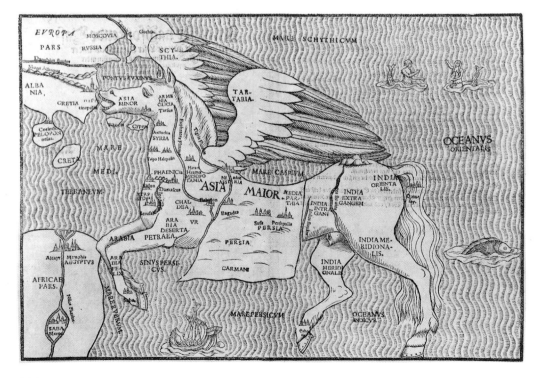

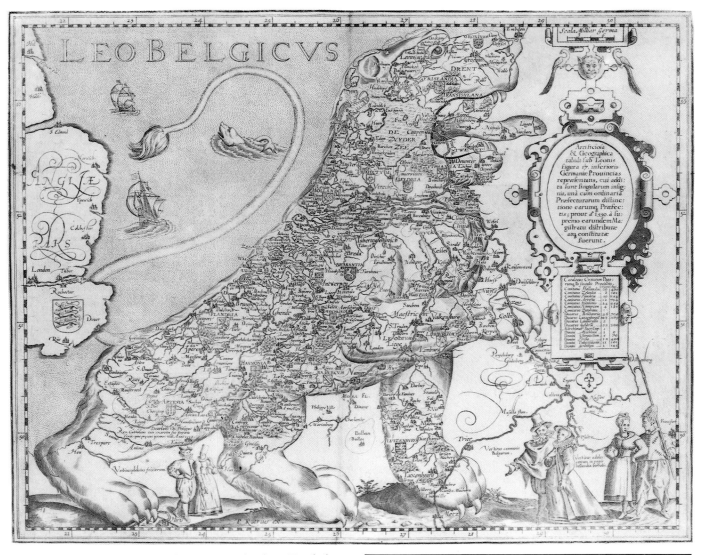

ally. During the eighteenth century further English, French, Dutch and Italian map makers issued sets of playing cards showing countries of the world.

The combination of maps with games was repeated in France in 1645 with the publication, by Pierre du Val, of *Le Jeu du Monde*, a simple dice board game in which each circular position has a map of a different country.

Over the next 200 years the educational element of such board games became paramount and the player would travel, for example, by dice throws or teetotum spins, the French Provinces, the counties of England and Wales, the countries of Europe or the World, and would identify, against a numerically keyed handbook, historical, physical or economic features relating to the particular place on which he, or she, had landed. This style of educational map game became increasingly popular and, particularly around the 1840s and 1850s, there were produced some most attractive, map-based board games which incorporated vignette views of major cities or sights of each county or country.

A variation of the geographical board game was the moralistic board game where a traveller, as in Pilgrim's

Above
Leo Belgicus

Pieter van den Keere's renowned version of Michael Aitzinger's concept of the Lion of Belgium.
Amsterdam 1617
45 x 37cm 17.5 x 14.5in

Michael Aitzinger's novel design was first printed in 1583 and was copied by many of the major Low Countries' engravers in various forms. Some versions are fabulously ornate, others more plain, but this is one of the most decorative forms. Aitzinger's concept arose from two factors – firstly, the shape of the coastline and the borders of the Netherlands, Belgium and Luxembourg, and secondly, the fact that most provinces within that area include a lion on their provincial coats of arms.

Progress, might encounter imagined moral or physical dangers as he progressed around the route mapped out. Such games were the forerunners of today's Snakes and Ladders and their basis on real or fictitious maps lends them a unique charm.

The moralistic board game derived from the many imaginary maps that were produced, either as separate

publications to amuse, or as book illustrations. More commonly seen in the latter form, this style of map first appeared in Sir Thomas More's *Utopia* of 1516. Maps illustrating 'Utopia', its German counterpart 'Cockaigne' or 'Schraffenlandt', *Pilgrim's Progress*, *Gulliver's Travels* and innumerable children's books can be found, although most of these are still thankfully intact with the original volumes. The map 'Schraffenlandt' can be found most easily since this was issued in

a number of atlases by Dutch and German map makers between about 1700 and 1750.

Another popularly produced format was the map of the 'Land of Love', which, occurring in a variety of forms, is an allegory of life and love and the possible dangers encountered on a voyage through life. Thus, the Matrimonial Map plots a path past the shores of indifference, the rocks of jealousy and the sands of lust to arrive, hopefully, at the island of everlasting

Right
Middlesex

A late issue of Robert Morden's playing
card series of maps.
London 1676-c1770
6 x 9cm 2.25 x 3.75in

*This rare curiosity is from a late edition of Robert Morden's set
of playing cards published first in 1676. Originally printed as the
King of Hearts, the suite-mark was removed when the
copperplates were reprinted as a pocket atlas nearly one hundred
years later. (Printed here larger than life size.)*

*Despite the novel purpose behind these miniature maps, many
of the counties of England and Wales, which are shown
individually in the series, were the first of those areas to show
roads.*

Left
Geography Bewitched ... Scotland

Bowles' and Carver's amusing caricature
map of Scotland.
London c1780
18 x 21cm 7 x 8.5in

*Intended only to amuse, without the element of satire often seen
in political caricatures, the series of 'Geography Bewitched' shows
jaunty characters designed in the forms of England and Wales,
Scotland and Ireland.*

*The example here, in typical bold early colour, is of a jocular
Scottish character. Such cartographical amusements were particularly popular around this time and cartographic invention
flourished in the production of geographical games, fictitious maps
and so on.*

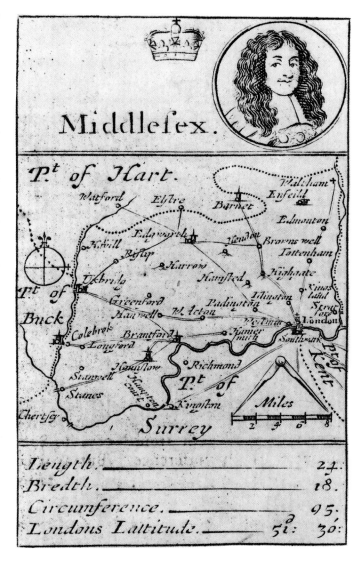

happiness as opposed to the Gulf of Scandal.

Maps produced solely for amusement, as in the above
case, include the political and social caricature maps
produced from about 1770 onwards. In many instances,
the purpose of such publications was satirical, but the
series of *Geography Bewitched* characters featuring the
home countries ran to many editions and was notable
for its lack of offence. During the nineteenth century,
the political satire and caricaturing standards set by

Gillray, Hogarth and others led to printed nationalistic
caricatures. Many involved the use of maps, either to
identify the nationality of the persons in question, or
to use that country's outline shaped into human form.
From around 1800, these political broadsheet cartoons
frequently illustrated the conflicts between England and
France. Later the style evolved to show a map of Europe
with each nationality represented by a caricature in
human or animal form.

Bibliography

Amongst many others, the following list of books, all currently in print, will indicate some of the further sources available to the student of the history of cartography or map collecting enthusiast. I have listed those titles of most general interest, reluctantly leaving unmentioned many others of equal merit but of more limited appeal.

★ ★ ★

The Map Collector Issued quarterly, this is the only international/national magazine for collectors.

Imago Mundi An annual publication comprising scholarly papers, research details and carto-bibliographic essays.

Maps and Map-makers R V Tooley's important general work, in print for the last forty years, an indispensable reference.

Antique Maps C Moreland and D Bannister – an excellent general reference book.

Tooley's Dictionary of Mapmakers R V Tooley's dictionary, with its supplement, is the only exhaustive listing of those associated with all aspects of map making and publishing.

Cartographic Innovations H W Wallis and A H Robinson. A scholarly investigation into mapping terms and practices.

The History of Cartography: Volume One J B Harley and D Woodward. The first of a five-volume definitive history of cartography through the ages. This volume extends only to medieval cartography.

The Earliest Printed Maps 1472-1500 T Campbell's detailed analysis of the 'incunabula' period of map printing.

Mapping of the World R W Shirley's definitive work on World maps printed before 1700.

The City in Maps J Elliot's summary of the history of urban cartography.

Early Printed Maps of the British Isles R W Shirley's listing and descriptions of maps printed before 1650.

Carto-bibliographies of some individual counties exist. However, the general works by T Chubb (covering 1579-1870), R A Skelton (1579-1703) and D Hodson (1703-1750) are of great use.

Looking at Old Maps and *Printed Maps of Wales* by J Booth illustrate and describe a good cross-section of interesting British and Welsh maps.

Victorian Maps of the British Isles by D Smith concentrates on a fascinating period of map development.

Printed Maps of London 1553-1850 by J Howgego and *Printed Maps of Victorian London 1851-1900* by R Hyde provide an exhaustive listing of maps of the capital.

The Early Maps of Scotland: Volumes 1 and 2 by the Royal Scottish Geographical Society.

Atlantes Neerlandici Professor C Koeman's immense carto-bibliography of atlases published in the Low Countries from earliest times to 1940.

Les Atlas Français XVI-XVII Siècles Madame Pastoureau's carto-bibliography of French-produced atlases.

Maps of Africa Dr O Norwich's listing and description of the major maps of the continent and its ports.

Arabia in Early Maps G R Tibbett's listing up to 1751.

Maps of the Bible Lands K Nebenzahl's finely illustrated survey of Holy Land maps.

India Within the Ganges S Gole's listing to 1800 of the development of maps of the sub-continent.

Early Maps of South East Asia R T Fell's summary covering the whole area.

Philippine Cartography C Quirino's detailed history and listing of maps of the island group.

Isles of Gold Sir H Cortazzi's summary of European and Japanese mapping of Japan.

Australia Unveiled Professor G Schilder's detailed history of early Dutch cartography in the Austral regions.

The Mapping of America S Schwartz and R Ehrenburg's exhaustive overall survey of North American map making.

The Mapping of America R V Tooley's compilation of previously issued papers in specifically interesting topics.

Maps of Texas and The South West 1513-1900 by J and R Martin is a well illustrated and described collection of some of the landmark maps of the region.

A Bibliography of printed Battle Plans of the America Revolution 1775-1795 by K Nedenzahl.

The Sky Explained D J Warner's analysis of printed star maps between 1500 and 1800.

Maps of the Heavens G S Snyder's beautifully presented commentary on manuscript and printed celestial charts.

Old Globes in the Netherlands P Van der Krogt's description of many of the major globes produced before 1800.

Cartographical Curiosities S G Hill's fascinating British Library exhibition catalogue.

Index